GRAVITY
AND THE EARTH

A. H. Cook, F.R.S. – University of Edinburgh

WYKEHAM PUBLICATIONS (LONDON) LTD
(A subsidiary of Taylor & Francis Ltd)
LONDON & WINCHESTER
1969

First published 1969 by Wykeham Publications (London) Ltd.

Cover illustration – Manned spacecraft view of the earth – by courtesy of Keystone Press Agency Ltd.

85109 070 2

Printed in Great Britain by Taylor & Francis Ltd.
Cannon House, Macklin Street, London, W.C.2

Worldwide distribution, excluding the Western Hemisphere, India, Pakistan and Japan, by Associated Book Publishers, Ltd., London and Andover.

FOREWORD

THIS book deals with the main facts about the gravity field of the Earth and the deductions that can be made about the internal state of the Earth. The subject was almost the first to be studied in an exact way in modern science and the theoretical side is very well developed. Since the subject lends itself to mathematical treatment, a good account cannot be given without mathematics. The algebra is not well covered in A-level courses but it is hoped that in the simplified form given in the book, it will be of interest to mathematicians and mathematical physicists and not too difficult for other physicists. The applications in geological exploration are of great importance for Earth scientists.

Despite the rather complete state of theory, the subject has been revitalized by two great technical developments. One, well known, is the launching of artificial satellites ; the other, less generally familiar, is the development of ways of measuring gravity at sea from ships. As a result, there is now far more interest, both fundamental and applied, in the gravity field of the Earth, than there has ever previously been. Now, quite recently, the launching of lunar satellites and space probes has made it possible to study the gravity fields of the Moon and eventually the planets, in some detail and in this way to extend our knowledge of the Moon and the planets, of their relations to the Earth, and of their origins. This is a good time to write a book on the gravity field of the Earth, for much new information has been obtained about the Earth in recent years and a knowledge of this will lead to an understanding of what lunar and planetary research will reveal in the near future.

I am most grateful to the late Mr. V. T. Saunders for careful reading of the manuscript and for many helpful suggestions. A. H. COOK

LIST OF SYMBOLS

There are certain special meanings in Chapter 6, as indicated.

A	moment of inertia about axis in equatorial plane.
a	radius; semi-major axis of ellipse or of meridian.
b	semi-minor axis of ellipse; polar axis of Earth.
C	moment of inertia about polar axis.
D	distance.
E	mass of Earth (Chapter 6).
e	eccentricity of ellipse.
f	polar flattening of Earth, $(a-b)/a$.
G	constant of gravitation.
g	acceleration due to gravity.
H	height; dynamic ellipticity of Earth, $(C-A)/C$.
h	height; Love's number (Chapter 6).
I	moment of inertia.
i	inclination of orbit.
J_n	coefficient of spherical harmonic in potential of Earth.
k	any constant; Love's number (Chapter 6).
L	length.
l	length; Love's number (Chapter 6).
M	mass; mass of Moon.
m	mass.
$\mathbf{m}$	ratio of centrifugal force to gravitational force at the equator (Chapter 5).
N	height of geoid above spheroid.
p	a perpendicular distance.
r	radius; radial co-ordinate.
T	period of pendulum or orbit.
t	time; thickness.
u	co-ordinate of satellite in orbit.
V	potential.
v	velocity.
x, y, z	Cartesian co-ordinates.
β	coefficient in formula for gravity.
λ	wavelength of light.
μ	gravitational mass, equal to the product GM.
ν	angular frequency.
ρ, σ	density.
χ, θ, ϕ, ψ	angles, angular co-ordinates.
ω	angular velocity of Earth; longitude of perigee of orbit.
☊	longitude of node of orbit.

CONTENTS

NUMERICAL VALUES

Constant of gravitation, G	$6{\cdot}67 \times 10^{-11}$ N m^2/kg^2.
Equatorial radius of Earth, a	6378 km.
Polar flattening of Earth, f	1/298·25.
Approximate value of g	9·8 m/s^2.

CHAPTER 1
introduction

Gravity is, of all forces, the one of which we are most aware, for we are continually subject to it, and many times a day we make use of it or find it an encumbrance. If our encounters with magnetic or electric forces are more striking, they are in nature somewhat exceptional; gravity, on the other hand, is part of the framework in which our lives are set. On the larger scale of the Universe, it is the same—the spectacular phenomena are the manifestations of electric or magnetic forces, but the force that keeps the planets in their courses around the Sun, or the stars in their paths in the galaxies, is that of the attraction of gravity. At the same time, gravity is a rather mysterious force. We know in detail how electrical and magnetic forces are related to the elementary electrical properties of electrons and atoms and to the consequences of relativity but the gravitational properties of the elementary particles of matter do not appear to be related to any of their other properties. One reason for our poor understanding of gravity is that it is an exceedingly weak force. Anyone climbing a mountain or lifting a heavy weight, or on the other hand, sailing down a ski run, may think that an inappropriate remark, but the reason that gravity seems so important to us is that the mass of the Earth is very great. A more fundamental observation is that the gravitational attraction between a proton and an electron is equal to the electrical attraction divided by 2×10^{39}—the lesson that we should draw from this comparison is that the Earth is electrically neutral to an exceedingly high degree since it has only an unimportant electrical field.

This book, however, is concerned with the gravity field of the Earth and although at the end something will be said about experiments on gravitation, the problems of the origin of gravitation and its relation to the geometry of the Universe, are left to books on relativity; the aim of this one is literally, more down to earth—it is to describe the measurement of the force of gravity produced by the Earth and how our knowledge of this force may lead us to understand the structure of the Earth.

We make use of gravity in a great many ways. Biology, engineering, the way we live in almost every respect, would be quite different if we were not attached to the surface of the Earth and given a way up, a head and feet, by the attraction of gravity. In our laboratories, we make daily use of the force of gravity to compare masses on the chemical balance, while the force of gravity acting on a column of mercury gives us the means of measuring the pressure of gases in a manometer or barometer. Our unit of electrical current depends on the force of gravity, for we establish it by comparing the force between two coils carrying currents with the force exerted by gravity on a known mass. We make use of gravity to study the make-up of the Earth, both the local geological structure near the surface of the Earth, as well as the nature of the Earth at great depths ; we put that knowledge to practical use in the search for oil and other minerals, while our detailed knowledge of gravity outside the Earth, the Moon and other planets is employed in the planning of space flights.

Although the gravitational attraction between two bodies is a force, we observe its effects in the form of the accelerations of one body towards another, and when we measure gravity on the surface of the Earth, it is the acceleration of a freely falling body towards the Earth that we measure. The units in which we express gravity at the surface of the Earth are therefore those of acceleration—metres per second per second or m/s^2 in the Système International d'Unités (SI) version of the metric system. The value of gravity at the surface of the Earth is very nearly 10 m/s^2—in England it is about 9·81 m/s^2—and gravity varies by 1 part in 200 from the equator to the poles. All other variations with height and with geological structure are much smaller, usually not more than 0·001 m/s^2, and sensitive instruments can measure changes as small as 0·000 000 1 m/s^2. The m/s^2 is thus an inconveniently large unit for stating the variations of gravity of geophysical interest and it is convenient to use a much smaller unit, the *milligal*, which is 10^{-5} m/s^2. The recognized abbreviation is *mgal*; the value of gravity in England is some 981 000 mgal, the variation from the equator to the poles is about 5000 mgal ; and variations of geological interest commonly lie between 1 and 100 mgal.

If two bodies, of negligible size but of masses M and m, are separated by a distance r, the force each exerts on the other is :

$$\frac{GMm}{r^2},$$

where G is the Newtonian constant of gravitation, about $6{\cdot}67 \times 10^{-11}$ $m^3/kg\ s^2$. The acceleration of the mass m towards M is thus :

$$\frac{GM}{r^2}.$$

Imagine a particle of unit mass to be a very great distance from the mass M and released so that it moves freely towards M; the work done on it in moving it to a distance r will be equal to the integral of the product of the mass times the acceleration over the distance up to that point, namely :

$$\int_{\infty}^{r} \frac{GM\,dr}{r^2} = -\frac{GM}{r}.$$

The quantity $-GM/r$ is called the *potential* of the mass M at the distance r; the potential is often denoted by V.

Suppose that the acceleration of a body of unit mass is g; the work done in traversing a distance dr in the direction in which gravity is acting is $g\,dr$, which must be equal to the change in potential:

$$-dV = g\,dr,$$

or

$$g = -\frac{dV}{dr}.$$

The negative sign comes in because in the neighbourhood of a point mass, gravity increases when the distance r from the mass decreases.

More generally, if the unit mass is moved in a direction that makes an angle θ with the direction of gravity, so that the component of the force acting in the direction of movement is $g \cos \theta$,

$$-dV = g \cos \theta \, dr,$$

and we say that the force of gravity is the *gradient* of a potential, the gradient being the maximum value of the differential $-dV/dr$ for changes of the direction of r. It is a vector quantity with a direction the same as that in which the rate of change of V is a maximum; in Cartesian coordinates, it has the components:

$$-\frac{\partial V}{\partial x}, \quad -\frac{\partial V}{\partial y}, \quad -\frac{\partial V}{\partial z},$$

along the directions of the coordinate axes.

It is often much easier to deal with the potential of a body rather than with the force of gravity because gravity is a vector and the resultant attraction of a set of masses has to be found by vector addition of the attractions of the individual masses, whereas the potential is a scalar and the potential of a composite body is found by scalar addition of the potentials of the components.

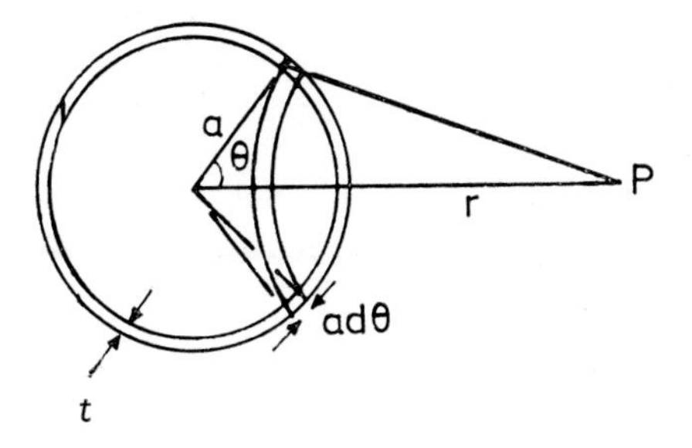

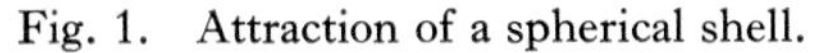
Fig. 1. Attraction of a spherical shell.

Consider, for example, the attraction of a spherical shell (fig. 1). Let the radius of the shell be a, let its thickness be t (taken to be very small) let the density of the shell be ρ and let a point P be at a distance r from the centre of the shell (r is greater than a). Around the line joining P to the centre of the sphere, draw cones with semi-angles θ and $\theta + d\theta$ and consider the potential due to the ring of the shell cut off by these two cones. The width of the ring is $a\,d\theta$, its thickness is t, its radius is $a \sin\theta$ and its circumference is $2\pi a \sin\theta$, so that its volume is :

$$2\pi a^2 t \sin\theta\, d\theta,$$

and its mass is :

$$2\pi\rho a^2 t \sin\theta\, d\theta.$$

Since every point lies at the same distance $(a^2 + r^2 - 2ar\cos\theta)^{\frac{1}{2}}$ from P, the potential of the ring is equal to its mass multiplied by G and divided by that distance, that is to say

$$dV = -\frac{2\pi G\rho a^2 t \sin\theta\, d\theta}{(a^2 + r^2 - 2ar\cos\theta)^{\frac{1}{2}}}.$$

The potential of the whole shell is obtained by integrating over the whole range of θ, namely from 0 to π :

$$\begin{aligned} V &= -2\pi G\rho a^2 t \int_{\infty}^{\pi} \frac{\sin\theta\, d\theta}{(a^2 + r^2 - 2ar\cos\theta)^{\frac{1}{2}}} \\ &= \frac{4\pi G\rho a^2 t}{r}. \end{aligned}$$

But the mass of the shell is $4\pi\rho a^2 t$ and so the potential of the shell is just the same as that of the same mass at the centre.

This is a most important result in the study of the Earth. If a spherical body is made up of a succession of spherical shells all having the same centre, then no matter what the density and therefore the mass of each individual shell, the potential of each will be the same as that due to its mass concentrated at the common centre, and therefore the potential of the whole body will also be equal to that of its total mass at the centre. The Earth is very close to having spherical symmetry, that is to say, its properties depend only on the radius from the centre, and the gravitational field of the Earth is very close to that of a point mass at the centre. As will be seen in Chapter 5, it is only the small departures from spherical symmetry that allow anything to be learnt from gravity about the internal structure of the Earth.

The plan of this book is as follows. Chapter 2 explains the methods available for the measurement of gravity at the surface of the Earth and Chapter 3 shows how the results of such measurements may be applied to the study of the geological structure of the upper parts of the Earth. Surface measurements give a poor view of the world-wide behaviour of gravity and in Chapter 4 it is shown how the study of the paths of artificial satellites of the Earth enables a comprehensive overall view to be obtained, while in Chapter 5 this wider knowledge is applied to the study of the deep structure of the Earth, supplemented, in Chapter 6, by the information derived from the way in which gravity varies with the different positions of the Sun and the Moon. Finally, in Chapter 7, some experiments on the constant of gravitation are described. No references are given in the chapters; but at the end of the book, there are suggestions for further reading.

CHAPTER 2

the measurement of gravity

WHEN we speak of the measurement of gravity, we mean the measurement of the acceleration of a body moving under no force except the gravitational attraction of the Earth. The most accurate and most straightforward way of measuring the acceleration due to gravity at the present time is in fact to measure directly the acceleration of such a freely moving body, but although this may seem the obvious thing to do, it is only quite recently that it has been possible to make the measurements with sufficient accuracy. There have been two big difficulties in the way of making direct measurements of gravity. It is convenient in a laboratory to let a body fall through a distance of about 1 m, which will take about 0·45 s. It has long been possible to measure a distance of 1 m to about one part in a million, but it was only since the development of electronic timing systems for radar that it became possible to measure a time of half a second to one part in a million, that is to half a microsecond. Until then, gravity was measured with pendulums, using the relation that the periodic time, T, of a pendulum is related to the length, l, by the formula :

$$T^2 = 4\pi^2 l/g.$$

Again, the length of the pendulum may be about 1 m, and the period about 1 s, but because the pendulum can swing for a few hours, a measurement of the total time of swing accurate to one-hundredth of a second will, when combined with a count of the total number of oscillations, give the period correct to one part in a million.

Absolute measurements of gravity

Imagine a body falling freely under gravity and suppose that it passes a height x at a time t. The height and the time are both measured from arbitrary origins. Then x and t are related by the formula :

$$x = a + ut + \tfrac{1}{2}gt^2,$$

where a and u are arbitrary constants and g is the acceleration due to gravity. a is the distance of the body from the origin of height at the instant from which the time is measured and u is the velocity of the body at that instant. The measurements are always arranged so that it is not necessary to know a and u to begin with. The simplest scheme is to measure the times at which the body passes three heights. Let these be at x_1, x_2, x_3 and let the measured times be t_1, t_2, and t_3. Then

$$
\begin{aligned}
x_1 &= a+ut_1+\tfrac{1}{2}gt_1^2,\\
x_2 &= a+ut_2+\tfrac{1}{2}gt_2^2,\\
x_3 &= a+ut_3+\tfrac{1}{2}gt_3^2,
\end{aligned}
$$

and therefore

$$
\begin{aligned}
x_1-x_2 &= (t_1-t_2)[u+\tfrac{1}{2}g(t_1+t_2)],\\
x_1-x_3 &= (t_1-t_3)[u+\tfrac{1}{2}g(t_1+t_3)],
\end{aligned}
$$

that is

$$
\begin{aligned}
\frac{x_1-x_2}{t_1-t_2} &= u+\tfrac{1}{2}g(t_1+t_2),\\
\frac{x_1-x_3}{t_1-t_3} &= u+\tfrac{1}{2}g(t_1+t_3),
\end{aligned}
$$

so that

$$\tfrac{1}{2}g(t_2-t_3) = \frac{x_1-x_2}{t_1-t_2}-\frac{x_1-x_3}{t_1-t_3},$$

or

$$g = \frac{2\{(x_1-x_2)(t_1-t_3)-(x_1-x_3)(t_1-t_2)\}}{(t_1-t_2)(t_1-t\ )(t_2-t_3)}.$$

The first accurate measurements with a falling body were made by photographing the lines on a falling graduated scale. The bar was lit by a series of very short bright flashes of light at known intervals of time as it passed a fixed camera. Because it was convenient to use the large number of lines available on a graduated scale instead of just three, and also in order to increase the precision of the result by making

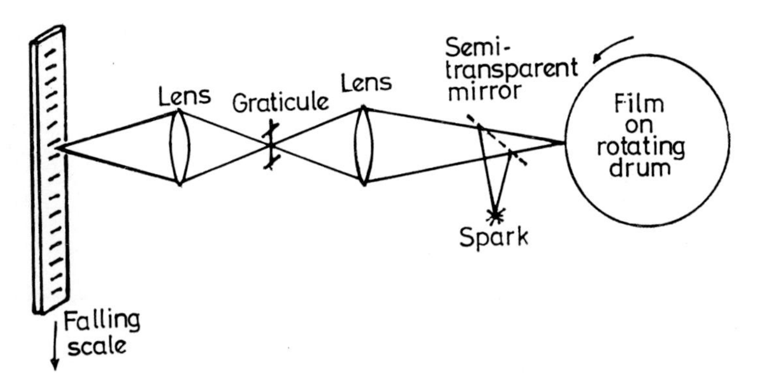

Fig. 2. Free fall determination of gravity (by Å. Thulin at the International Bureau of Weights and Measures, Sèvres).

more observations, some fifty pairs of distances and times were observed instead of the three that are the necessary minimum. A diagram of the apparatus is shown in fig. 2. The scale was 1 m long and one face of it was highly polished and had engraved on it narrow lines at 1 mm intervals. The face of the bar was imaged by the lens of a camera on to a moving strip of film on which the image of a fixed graticule was also projected. The face of the bar was illuminated by a spark of high intensity lasting for only a few millionths of a second; the spark was driven by an electrical voltage at accurately known intervals of time so that the instant of each photograph of the face of the bar was known to about one millionth of a second. The position of the centre of mass of the bar at the time of each photograph could be worked out from the number of the graduation mark that was photographed, together with the distance between the image of the mark and that of the fixed graticule. The accuracy of this experiment was somewhat better than one part in a million (1 mgal).

A very much more accurate determination has recently been made, using the same principle of free-fall but with an entirely different means of measurement. If light which is highly monochromatic, especially that given out by a gas laser, falls on a plate that is semi-transparent and reflects some of the light and transmits another part, and if the transmitted and reflected beams, after being reflected by mirrors as shown in fig. 3, are directed by the semi-reflecting mirror to a detector, the output of the detector will vary with the positions of the two mirrors, that is the light will *interfere* at the detector. If the light travels a distance d further between the source and the detector when

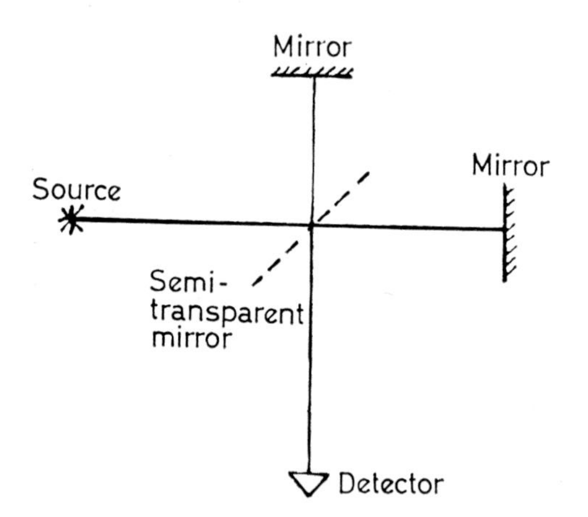

Fig. 3. Principle of the Michelson interferometer.

it goes by one mirror as compared with the other, the output of the detector will be proportional to :

$$1 + V \cos 2kd,$$

where V is a constant, less than 1, which expresses the contrast of the interference effect, and k is equal to $2\pi/\lambda$, λ being the wavelength of the light—about $0 \cdot 6$ μm for the red light from a helium–neon laser. If now one of the mirrors is moved, the interference can be used to find out how far it has gone, for all that is necessary is to count the number of times that the output of the detector passes through a maximum and that will be equal to four times the number of wavelengths of light in the distance moved. Electronic counters are available that can count the number of peaks at a rate corresponding to the speed of a body after it has fallen freely for 1 m, namely 30 MHz, or thirty million times a second.

This simple interferometer cannot be used directly for the measurement of gravity. The various mirrors must be adjusted so that the two reflected beams exactly coincide when they arrive at the detector, or no interference will occur. Were one of the mirrors falling, it would not remain parallel to itself and so the beam reflected from it would not always arrive at the same part of the detector and interference would not be obtained. However, there are reflector systems that always return a beam of light parallel to itself, and if one such is dropped, the light reflected from it will always coincide on the detector with that reflected from a fixed mirror. One such reflector system is the combination of lens and mirror shown in fig. 4. N is

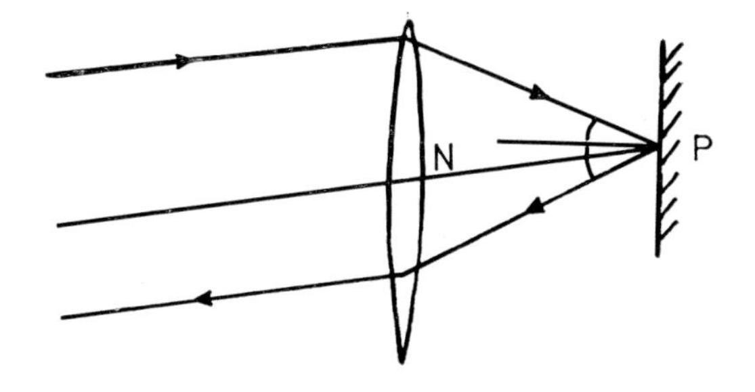

Fig. 4. Cat's-eye reflector.

the nodal point of the lens, M is the mirror placed in the focal plane of the lens, and P is the point at which a ray through N meets the mirror. By the focal property of the lens, any incident ray parallel to PN passes through P and any ray originating from P is parallel to PN after traversing the lens. In consequence all incident rays parallel to PN come to a focus at P and after reflection at the mirror, re-emerge from the lens parallel to *PN*. Another system comprises three mirrors mutually at right angles—a cube-corner reflector—of which a two-dimensional analogue is shown in fig. 5.

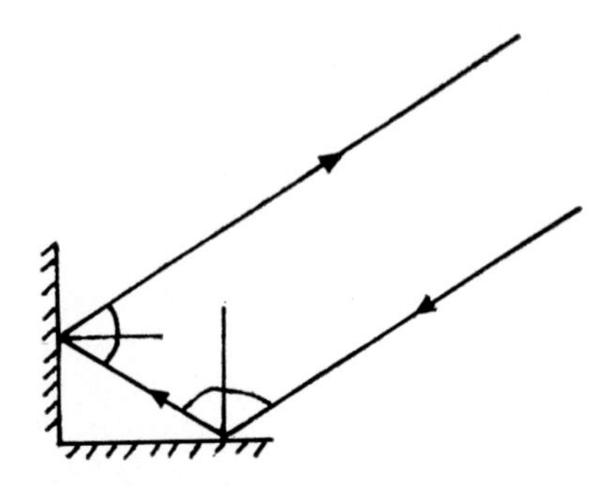

Fig. 5. Cube-corner reflector.

A diagram of apparatus for measuring gravity using a lens-mirror combination is shown in fig. 6. An interferometer is formed of the lens-mirror, a semi-reflecting mirror, and a fixed mirror. The lens-mirror can fall freely in a tube in which a high vacuum is maintained. Light from a gas laser illuminates the interferometer and is detected by a photocell, the output from which goes to a pair of electronic

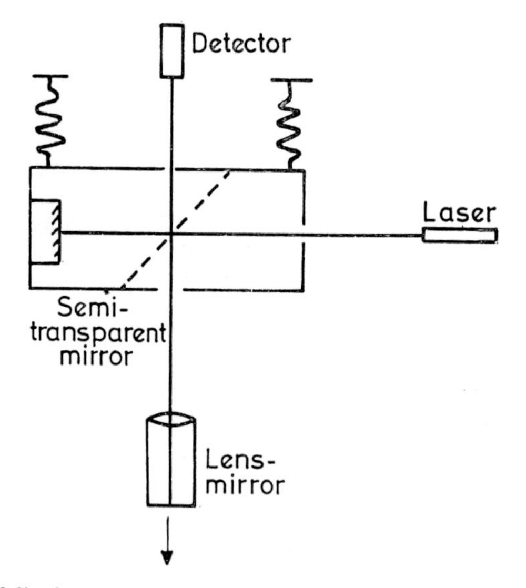

Fig. 6. Free-fall determination of gravity with an interferometer (J. E. Faller, Wesleyan University, Connecticut).

counters pre-set to count different numbers of interference peaks. Two other counters count the numbers of periods of an electrical signal of known frequency that correspond to the times taken for the lens-mirror to fall through the two distances given by the numbers of counts of fringes accumulated in the first two counters. In order that a large number of observations can be made easily, it is arranged that the lens-mirror is caught at the bottom of its fall in a holder which can then be raised by an electric motor controlled from outside the vacuum vessel to return the lens-mirror to the top of the tube where it is held magnetically until it is to be dropped again.

One problem with all measurements of gravity involving the direct observation of a freely falling body is the movement of the apparatus by which the fall is observed. The value of the acceleration that is required is that relative to the centre of mass of the Earth but the acceleration observed is that relative to the laboratory apparatus. Now the ground is always in motion, due in part to the effect of distant sea-waves and in part to the effect of machinery, traffic and similar local causes. The former motions have amplitudes of about 1 or 2 μm (millionths of a metre) and periods of 4 to 7 s, while the latter have amplitudes of about 0·1 μm and periods of a few tenths of a second. In the apparatus shown in fig. 6, the fixed mirror and the semi-reflecting mirror of the interferometer are placed on a platform that is not rigidly fixed to the ground but is supported on a system of springs with a very long period of oscillation that isolates the mirrors from the movements of the ground; in this way the measured acceleration is that relative to the centre of mass of the Earth.

Another problem with a falling body is the reduction in the acceleration due to the resistance of the air. Because the viscosity of a gas is independent of pressure, the resistance is constant at pressures above about one-thousandth of an atmosphere, but at lower pressures, where the air can no longer be thought of as a fluid but rather as a cloud of separate molecules, the resistance falls away, but even so amounts to nearly ten millionths of the gravitational force when the pressure is a millionth of an atmosphere. The pressure of air in which the body falls should therefore be less than one thousand-millionth of an atmosphere if a value of gravity correct to one part in a hundred million is to be obtained from the measurement. It is not difficult to produce such low pressures with modern pumps, and since the wavelength of the light emitted by the laser and the frequency of the electrical signal used for timing can both be established to better than one part in a hundred million, this laser interferometer apparatus should allow gravity to be measured to that accuracy.

Symmetrical free-motion measurements

A somewhat different arrangement of the experiment gives a result that is unaffected by the resistance of the air, at least provided the pressure of air is less than about one ten-thousandth of an atmosphere. The principle is shown in fig. 7. An object is thrown up vertically and is timed as it goes up and as it falls back again. Let the

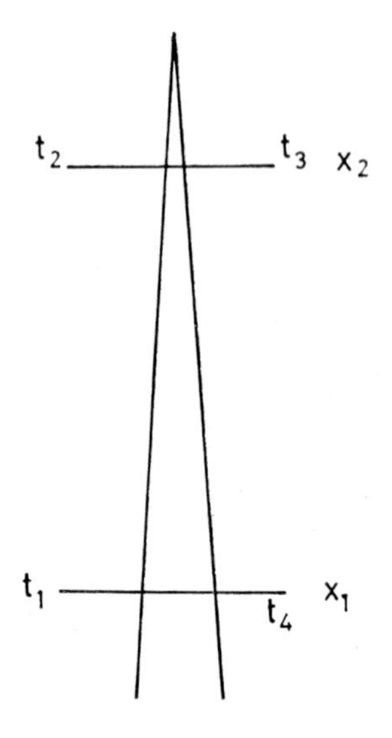

Fig. 7. Principle of symmetrical free-motion determination of gravity.

times be measured at two heights, x_1 and x_2. Suppose that the time at which the body passes height x_1 on the upward flight is t_1 and that at which it passes height x_2 be t_2. Let the times on the downward path be t_3 at x_2 and t_4 at x_1. Then

$$x_1 = a+ut_1+\tfrac{1}{2}gt_1^2,$$
$$x_2 = a+ut_2+\tfrac{1}{2}gt_2^2,$$
$$x_2 = a+ut_3+\tfrac{1}{2}gt_3^2,$$

and

$$x_1 = a+ut_4+\tfrac{1}{2}gt_4^2.$$

Hence

$$u(t_4-t_1) = -\tfrac{1}{2}g(t_4^2-t_1^2),$$

or

$$u = -\tfrac{1}{2}g(t_4+t_1).$$

But, also

$$u = -\tfrac{1}{2}g(t_3+t_2).$$

Thus, the mean values of the times at x_1 and x_3, $\frac{1}{2}(t_4+t_1)$ and $\frac{1}{2}(t_3+t_2)$ respectively, are equal.

Let the common mean value be t and let

$$t_4-t_1 = T_1, \quad t_3-t_2 = T_2.$$

Then

$$t_1 = t - \tfrac{1}{2}T_1, \quad t_4 = t + \tfrac{1}{2}T_1,$$
$$t_2 = t - \tfrac{1}{2}T_2, \quad t_3 = t + \tfrac{1}{2}T_2.$$

Also

$$\begin{aligned} x_2 - x_1 &= u(t_2 - t_1) + \tfrac{1}{2}g(t_2^2 - t_1^2) \\ &= (t_2 - t_1)[-\tfrac{1}{2}g(t_3 + t_2) + \tfrac{1}{2}g(t_2 + t_1)] \\ &= \tfrac{1}{2}g(t_2 - t_1)(t_1 - t_3) \\ &= -\tfrac{1}{2}g \,.\, \tfrac{1}{2}(T_1 - T_2)\tfrac{1}{2}(T_1 + T_2). \end{aligned}$$

Put $x_1 - x_2 = H$.

Then

$$g = \frac{8H}{T_1^2 - T_2^2}.$$

This is the formula for free motion under gravity alone. Now suppose that a resistive force proportional to the velocity is acting. If that force is $k\dot{x}$, the net acceleration, $\dot{x}$, of a body moving downwards is :

$$\ddot{x} = g - k\dot{x}$$

(x increases downwards).

The solution of this equation with the initial conditions $\dot{x} = 0$, $x = 0$ at $t = 0$, is :

$$x = \frac{g}{k}\left[t - \frac{1}{k}\left\{(1 - \exp(-kt)\right\}\right].$$

When k is small,

$$x = \tfrac{1}{2}gt^2(1 - \tfrac{1}{3}kt).$$

Our aim, however, is to calculate the small extra time that it takes to fall the distance h from the point of release in the presence of the resistance. If k were zero, the time would be $(2h/g)^{\frac{1}{2}}$; in the presence of resistance let it be $(2h/g)^{\frac{1}{2}} + \tau_d$. It will then be found by substitution that

$$\tau_d = \frac{1}{3}\frac{kh}{g}.$$

We also require the change in time to rise a distance h when the body is projected upwards with a velocity just sufficient to take it to that height in the presence of resistance. That velocity will be somewhat greater than is required in the absence of resistance, and therefore the time taken will be somewhat less than without resistance. If the body is projected with a velocity $-v$ (upwards) from the level $x = h$ at time $t = 0$, the velocity required to take it to the level $x = 0$ is :

$$g\left(\frac{2h}{g}\right)^{\frac{1}{2}}\left\{1 + \tfrac{1}{2}k\left(\frac{2h}{g}\right)^{\frac{1}{2}}\right\} + g\tau_u,$$

where τ_u is the excess of the time over $(2h/g)^{\frac{1}{2}}$. τ_u is then found to be :

$$-\tfrac{1}{3}\frac{kh}{g}.$$

Since this is equal to but of opposite sign to τ_d, the total time to rise from and fall back to a given level is the same whether resistance is present or not. The reason is that the extra time to fall through a given height is balanced by the reduction in time to rise to that height on account of the greater initial velocity of projection.

A diagram of the apparatus used in the first up-and-down experiment, which was performed at the National Physical Laboratory, is shown in fig. 8. In all free-motion experiments it is essential that

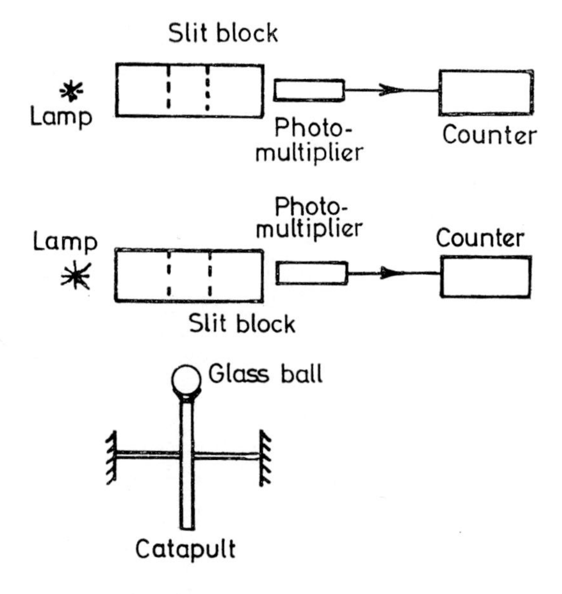

Fig. 8. Symmetrical free-motion experiment at the National Physical Laboratory, Teddington (A. H. Cook).

rotation of the moving object should have no effect on the measured acceleration ; the most certain way of ensuring this is to use a sphere. A glass sphere was used in the N.P.L. experiment and its position was detected by using it as a lens. It is of course, a very poor lens, but if it is placed midway between two slits at the correct separation, three times the diameter of the ball if the refractive index of the glass is 1·5, it will focus one slit quite sharply on the other, and if a light is placed behind one slit, then the light transmitted through the system will show a sharp maximum when the centre of the ball lies in the plane containing the slits. The arrangement is shown in fig. 9. In

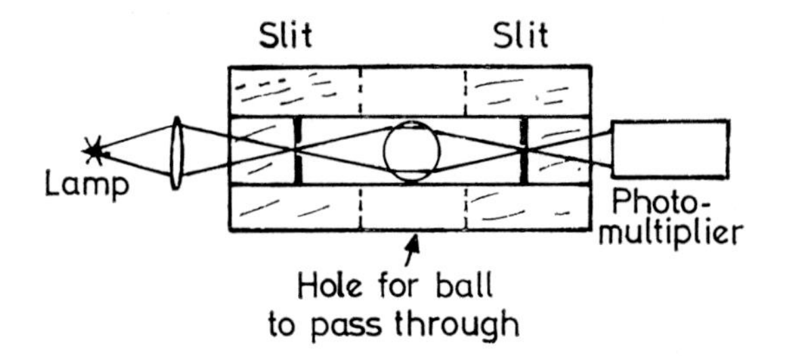

Fig. 9. Optical system of glass ball between two slits.

the N.P.L. experiment, it was possible to detect the passage of the ball across the plane of the slits to within about 0·1 μm, corresponding to an uncertainty of 0·1 mgal in gravity. Two pairs of slits were required, one for the upper level at which the ball was timed, and one for the lower level. They were mounted on glass blocks with holes in them through which the ball could pass on its flight; these blocks had optically flat horizontal surfaces which formed parts of an interferometer by which the separation of the blocks was measured. It is not possible to say exactly where the effective plane of the slits lay in the glass blocks, but by turning the blocks over and repeating the experiments, the unknown position of the slits can be eliminated.

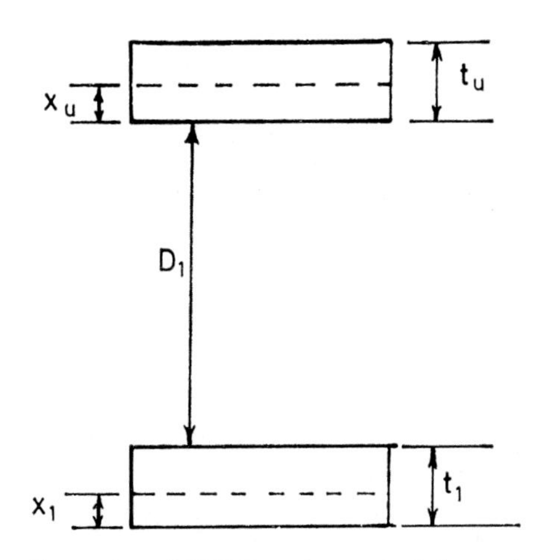

Fig. 10. Geometry of N.P.L. symmetrical determination of *g*.

The geometry is shown in fig. 10. Let D_1 be the measured separation of the blocks in the first experiment and let x_u be the position of the slit in the upper block above the lower face, and x_l that of the slit in the lower block above the lower face. Let t_u and t_l be the thicknesses of the blocks, measured interferometrically. The effective separation of the slits is then

$$H_1 = D_1 + x_u + t_l - x_l,$$

and if the measured times spent above the upper and lower planes are respectively T_u and T_l and if $\Delta T_1^2 = T_l^2 - T_u^2$, then

$$\tfrac{1}{8} g \Delta T_1^2 = H_1 = D_1 + x_u + t_l - x_l.$$

Now let both blocks be turned over. The effective separation of the slits becomes :

$$H_2 = D_2 + t_u - x_u + x_l,$$

and therefore

$$\tfrac{1}{8} g \Delta T_2^2 = D_2 + t_u - x_u + x_l.$$

Taking the mean of the two experiments,

$$\tfrac{1}{8} g \,.\, \tfrac{1}{2}(\Delta T_1^2 + \Delta T_2^2) = \tfrac{1}{2}(D_1 + D_2) + \tfrac{1}{2}(t_u + t_l),$$

in which x_u and x_l do not appear. In practice, many more observations than two are made.

The time intervals T_u and T_l were derived from photo-electric recording of the light coming through the slits. A photomultiplier placed behind the exit slit gives a sharp pulse of current when the ball is midway between the slits. The successive pulses were used to start and stop a counter counting the oscillations of an electrical signal of known frequency and in addition, the pulses were recorded on a cathode ray tube, the trace of which, with time markers superposed, was photographed. The counter gave the time interval to the nearest microsecond and the times of the peaks of the signals could be measured off the photographs to about one-tenth of a microsecond.

A difficulty with timing the simple fall of an object is that at the start of the fall, the object is moving rather slowly while it will be moving much more quickly at the end. The signals derived from the passage of an object past fixed levels will therefore last for a shorter time at the end of the fall than at the start and, in such circumstances, it is easy for systematic errors to occur in the measured time interval between signals at the start and at the end of the drop. In the up-and-down experiment, the signals between which the time intervals are measured occur at the same levels and thus at the same velocity of the moving object ; the signals are thus very closely of the same form and the chance of a systematic error is much less.

Experiments at different pressures of air confirmed the theoretical prediction that the measured acceleration should be independent of air resistance provided that the resistance is proportional to velocity. In fact, a significantly lower acceleration was only observed at about one-thousandth of an atmosphere, at which pressure the buoyancy of the air begins to affect the measurements. The definitive measurements were performed at much lower pressures, the exact values of which are not critical. The main source of error in the final result was due to the ground motion, no steps having been taken to eliminate or correct for it. By making about one hundred separate observations, the final uncertainty was reduced to less than 0·2 mgal.

A third recent measurement of gravity, and probably the most accurate of all, is that made at Sèvres at the International Bureau of Weights and Measures (fig. 11). This is an up-and-down experiment

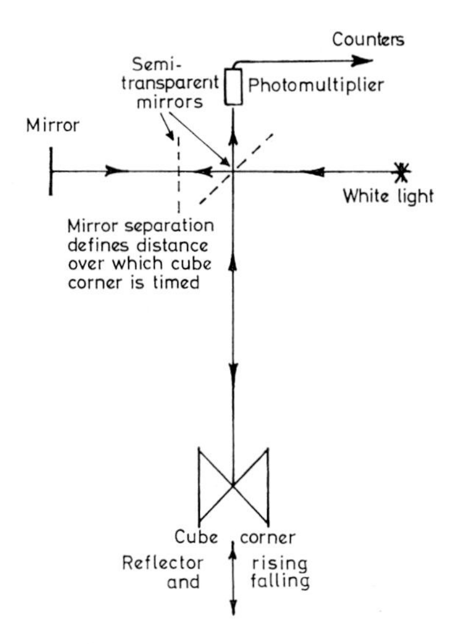

Fig. 11. Symmetrical free-motion determination of *g* at the International Bureau of Weights and Measures, Sèvres (A. Sakuma).

in which the moving object incorporates a cube-corner reflector (in fact it consists of two placed back-to-back to ensure that the centre of mass coincides with the optical centre of the system) and the two levels above which the times of flight of the body are measured are fixed by interferometric observations. Interference fringes are usually only seen if the light is nearly monochromatic but in some special arrangements fringes can be seen in white light. The system used in the experiment at Sèvres is shown in fig. 12. It is a Michelson interferometer in which the two paths from the beam divider D to the mirrors M_1 and M_2 are of equal length so that the path difference between the two beams is zero when they recombine. The phase shift introduced by the interferometer is therefore the same—zero—for all wavelengths, and so if white light passes through the interferometer, the emergent light has a greater intensity than for any other path difference. Figure 13 shows how the intensity of white light passing through such an interferometer varies with the path

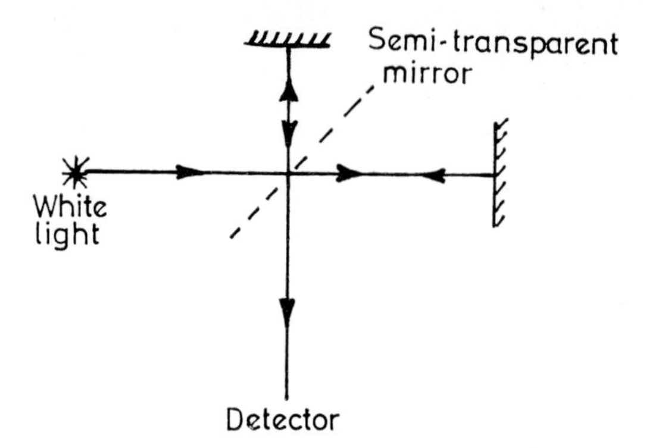

Fig. 12. White light interferometer.

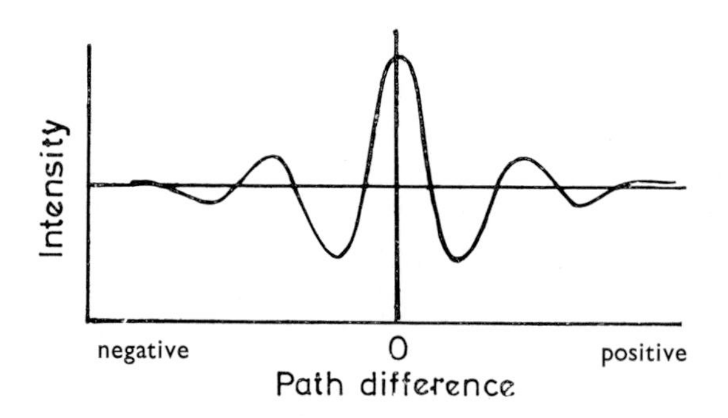

Fig. 13. Variation of intensity of light in a white light interferometer.

difference in the interferometer In the experiment at Sèvres, the cube-corner reflector is one of the mirrors of such a white light interferometer and when, in its free movement, it passes through the level corresponding to zero path difference in the interferometer, the white light, falling on a photomultiplier, gives a sharp electrical pulse that is used to start or stop a counter that counts the pulses of an electrical signal of known fixed frequency. By having two fixed mirrors in the interferometer, two levels are defined and the distance between them can be found from interferometric measurements of the separation of the two fixed mirrors. By suspending the whole interferometer on spring mounts that filter out movements of the ground, the measurements are made free of the effects of such movements and this, combined with the high precision of interferometric observations, has led to a much more accurate determination than the N.P.L. experiment.

In all free-motion gravity determinations, there is one type of disturbance that must be very carefully guarded against. The moving object has to be dropped or thrown up and in so doing there will be a reaction against the general framework of the apparatus that could lead to small mechanical oscillations of the framework. These oscillations, just like the natural motions of the ground, would cause errors in the measured value of gravity, but they would be much more serious because the effect would always be nearly the same for each separate flight of the moving object. The natural movement of the ground occurs randomly with the respect to the times at which the object passes the timing levels and so if enough separate observations are made, the average of them all will be nearly correct, as in the N.P.L. experiment, but the effect of a reaction of the launching of the object would not average out in that way. The reaction must therefore be investigated with great care to ensure that it does not disturb the measured acceleration.

The experiments made at the N.P.L. and at Sèvres used apparatus that could not easily be taken from place to place, but the Wesleyan (fig. 6) apparatus can be used in different places and has in fact been taken to both the N.P.L. and Sèvres, where comparative measurements have shown that the three methods agree to within 0·2 mgal.

Measurements of differences of gravity

The experiments so far described measure the value of gravity absolutely at the site at which they are performed—that is to say, they give the value of the acceleration in terms of the basic units of length and time. Such determinations are important fundamental measurements in physics and astronomy, as has already been indicated in the Introduction and as will be further discussed in Chapter 5; for the great majority of purposes, however, it is the relatively small variations of gravity from place to place that are important, both for prospecting for oil and minerals near the surface of the Earth and for understanding the deep structure of the Earth. Absolute determinations take much too long and are far too expensive for such purposes and indeed are not sufficiently sensitive. Variations of gravity are therefore measured with instruments specially designed for the purpose, quick to use and easy to carry so that many observations may be made rapidly even in places to which it is difficult to get. Such gravity meters almost all depend on suspending a mass with a spring and then observing the small changes in the position of the mass or the balancing force as the value of gravity changes.

Gravity meters

The principle of the great majority of gravity meters used on land is shown in fig. 14. A beam carrying a mass at one end is pivoted at the other and is maintained horizontal against the attraction of gravity by the tension of the spring. Let b be the length of the beam and let the tension in the spring be given by :

$$T = k(l - l_0)$$

where k is the spring constant, l is the extended length of the spring and l_0 is the length at which the tension is zero.

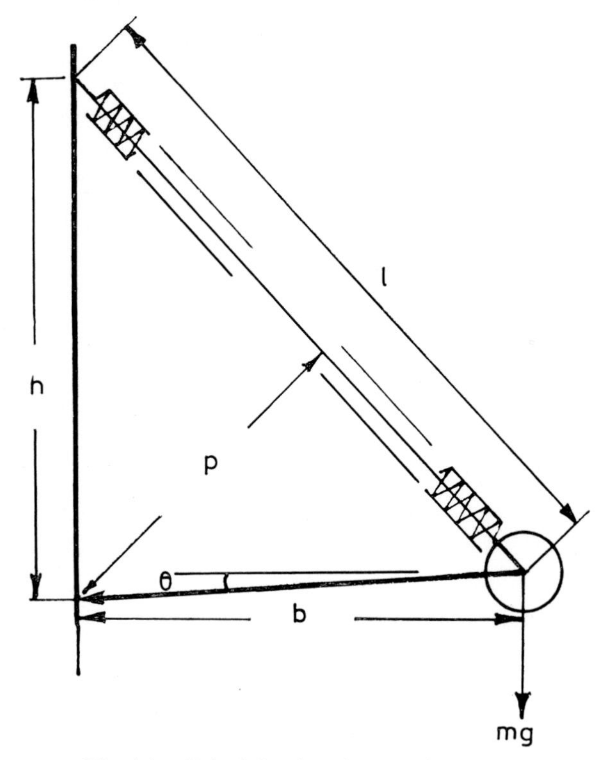

Fig. 14. Principle of spring gravity meter.

If h is the height of the point of attachment of the spring above the pivot of the beam, and if θ is the angle that the beam makes with the horizontal,

$$l^2 = h^2 + b^2 - 2hb \sin \theta,$$

while p, the length of the perpendicular from the pivot to the spring, is given by :

$$p = \frac{hb}{l} \cdot \cos \theta.$$

At equilibrium, the moments of the weight of the mass and of the tension of the spring about the pivot are equal :

$$pT = mgb \cos \theta \quad \text{or} \quad T = mgl/h.$$

But

$$T = k(l - l_0)$$

and therefore

$$l/l_0 = \frac{kh}{kh - mg}.$$

Hence

$$l = (h^2 + b^2 - 2hb \sin \theta)^{\frac{1}{2}} = \frac{khl_0}{kh - mg},$$

and since θ is nearly zero :

$$\frac{d\theta}{dg} = \frac{ml_0^2 k^2 h}{b(kh - mg)^3},$$

or

$$g\frac{d\theta}{dg} = \frac{l - l_0}{l_0}\left(\frac{h}{b} + \frac{b}{h}\right).$$

From this it can be seen that the sensitivity $d\theta/dg$, that is the change of angle of the beam for a given change of gravity, is greater the smaller the value of l_0; it also depends on the geometry but on no other factor. To make $d\theta/dg$ large, l_0 must be small.

l_0 can be made very small (the spring is then called a zero-length spring) by a special technique of winding the coils of a helical spring whereby the wire is twisted about an axis in its own length as it is wound.

In practical instruments a very delicate spring is attached one end to the beam near the mass and the other to a micrometer screw and by adjusting the micrometer the force on the main beam is altered so that when a change of gravity occurs, the beam is always returned to the same angle with the horizontal. The position of the beam is read by a sensitive optical device and it is commonly possible to detect changes of gravity as little as 0·01 mgal ($10^{-8}g$).

For such very delicate measurements, the spring constants k and l_0 must remain very stable. By differentiating the expression for l with respect to l_0 or k instead of with respect to g, it is found that

$$g\frac{\partial\theta}{\partial g} = -\,l_0\frac{\partial\theta}{\partial l_0}\left(\frac{l - l_0}{l_0}\right)$$

and

$$g\frac{\partial\theta}{\partial g} = -\,k\frac{\partial\theta}{\partial k}.$$

Thus if g is to be measured to 1 part in 10^8, k must be constant to 1 part in 10^8 and l_0 must be constant to an even higher degree. k varies with temperature, partly through the length of the spring changing but, much more important, through the change of elastic modulus with temperature, which for many materials is at the rate of 1 part in 10^4 per degree Kelvin. Special alloys of iron and nickel (Elinvar) have much smaller coefficients but even so, the gravity meter must be maintained at a very steady temperature. Either it is placed in an electrically controlled thermostat or it is made so small that it can be placed in a sealed vacuum flask. The Worden meter has a spring of fused silica which is very stable but which has the usual high temperature coefficient; by incorporating a metal element in the fine measuring spring, a large degree of temperature compensation is achieved and the whole instrument is placed in a sealed vacuum flask.

Fused silica is a very stable material and does not show a large change of l_0. Metal alloys are in general rather unstable, especially if they have been subject to the heat treatment and mechanical working necessary to produce the low rate of change of elastic modulus with temperature. However, the problem of producing a spring that shall have a small value of l_0, a low coefficient of temperature and a small drift has been intensively studied and, especially in the La Coste–Romberg gravity meter, has been in essence solved.

Using a gravity meter

Making a gravity survey with spring-balance gravity meters is a relatively simple matter. The main point that controls the work is that if gravity is to be measured to 1 part in 10^8, the position of the gravity meter in height even more than horizontally, must be very well known. Even in a well mapped country (and for this purpose there are few countries as well mapped as Britain), the determination of height to better than say 20 cm is quite difficult, especially when it is considered that the heights and positions of road surfaces are frequently altered. The main effort in a gravity survey has therefore to be devoted to finding the positions of the observations with sufficient accuracy, the actual readings of the gravity meter being quick and simple.

Pendulum measurements

In spring-balance gravity meters the change of gravity is proportional to the change in angle of the beam or to the rotation of the micrometer; these are not accelerations and have therefore to be calibrated in terms of a change of gravity. It has been found that this can only be done reliably by reading the meter at places at which gravity is already known. Ultimately, the measurements must be related to absolute measurements, but up to now very few absolute measurements have been made, and those that have are of much lower precision than the readings of a spring-balance gravity meter. Observations of pendulums have therefore been used to establish differences of gravity between a relatively small number of well-known accessible stations to which gravity meter observations can be referred. If I is the moment of inertia of a pendulum about its point of support and if h is the distance of the centre of mass below that point, the period T is given by :

$$T^2 = 4\pi^2 I/mgh.$$

If the mechanical properties of the pendulum do not change, the periods at two places at which the values of gravity are g_1 and g_2 are related by the expression :

$$\frac{T_1^2}{T_2^2} = \frac{g_2}{g_1};$$

$$\frac{g_2 - g_1}{g_1} = \frac{T_1^2 - T_2^2}{T_2^2}$$

Assuming that the absolute value of gravity at one place, g_1, say, is known, the difference between the two places can be calculated and used in the calibration of gravity meters. To measure the period of a pendulum, it is usual to reflect a beam of light from a mirror on the pendulum to a photocell so that whenever the pendulum passes through its position of rest, an electrical signal is derived from the photocell. By counting these pulses and the pulses from an electrical signal of known constant frequency, the period can be measured to a few parts in 10^8 in a time of about 1 hour.

The accuracy of measured values of differences of gravity is always worse and sometimes very much worse, than the accuracy of the measurement of the period at any one place would suggest ; in other words, the assumption that the mechanical properties of the pendulum stay constant when the pendulum is carried from one place to another is not borne out in practice. The problem has been given much attention in the last two decades, but cannot be said to be solved although some causes of the changes have been identified. The basic difficulty seems to lie in the support of the pendulum. The usual arrangement is to attach a block of hard steel worked to a very fine knife edge to the pendulum and to rest the edge on a very hard flat surface (fig. 15). The use of pendulums supported in this way for absolute gravity measurements has shown that the period of the pendulum can vary considerably with the position of the knife edge on the plane, no doubt because, however carefully the knife and plane are made, the contact between them will not be a uniform line but a series of spots where the high spots on the knife edge happen to meet those of the plane. The effective point of support will thus change from time to time, giving rise to changes in the period of the pendulum.

For this reason, the pendulum, in the well known reversible form introduced by Captain Henry Kater (1816), is no longer used for absolute measurements, and it is no doubt for the same reason that pendulum measurements of gravity differences are often disappointing.

None the less, the world-wide network of gravity observations depends on a few such pendulum observations, probably accurate to about 0·5 mgal, to tie together the more local surveys made with gravity meters and to establish common calibrations for the meters. Because gravity varies strongly with latitude, increasing from the

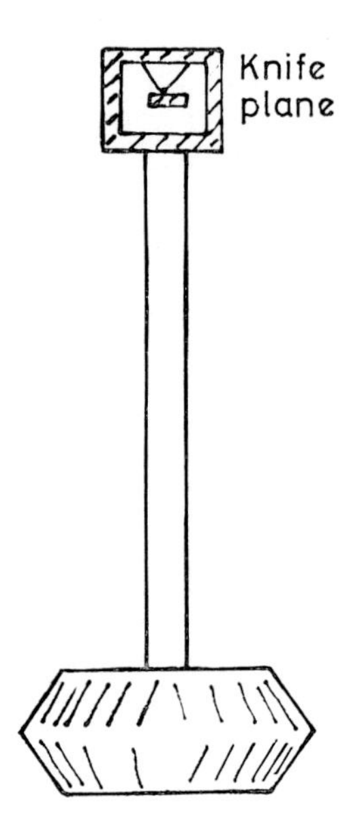

Fig. 15. Pendulum for measurement of differences of gravity.

equator to the poles, the necessary pendulum measurements are made along lines of roughly constant longitude. In Europe and Africa there is a set of sites of observations running from Cape Town through Johannesburg and Nairobi to Rome, Munich, Copenhagen and on to the north of Norway. Similarly, in the Americas, points lie on a line from Quito in Ecuador, through Mexico, to Denver, Alberta and on to Alaska. A similar line is established between Australia and Japan. Within the continents that are traversed by these lines, gravity surveys are well linked together by reference to the points on the lines, but the values in different groups of continents are less well connected.

Measurements of gravity at sea

The measurements that have been described so far all apply to values of gravity at places on the solid earth where the acceleration is constant, at least within the time taken for a single observation. They cannot therefore be made at sea where the actual acceleration of a point on a ship is the resultant of the acceleration due to gravity and the accelerations of the pitching, rolling and yawing motions of the ship under the action of the waves. Special methods must be applied to the measurement of gravity at sea, a very important thing to be able to do since the seas cover nearly three quarters of the surface of the Earth and gravity measurements are most important for understanding the structures of the sea basins.

Besides the oscillatory accelerations of the ship under the action of the waves, there is another acceleration arising from the steady east–west component of the motion of the ship. The observed acceleration of a falling body is the resultant of the gravitational attraction of the matter of the Earth and the acceleration experienced by a body moving in a circle with the angular velocity of the Earth and at the radius of the surface. Since the former is much the greater, the resultant is directed nearly toward the centre of the Earth and the component of the centrifugal acceleration away from the centre is

$$a\omega^2 \cos^2 \theta,$$

where a is the radius of the Earth, θ is the latitude, and ω is the angular velocity of the Earth (fig. 16). The value of this acceleration at the equator is 3400 mgal. The acceleration may alternatively be written

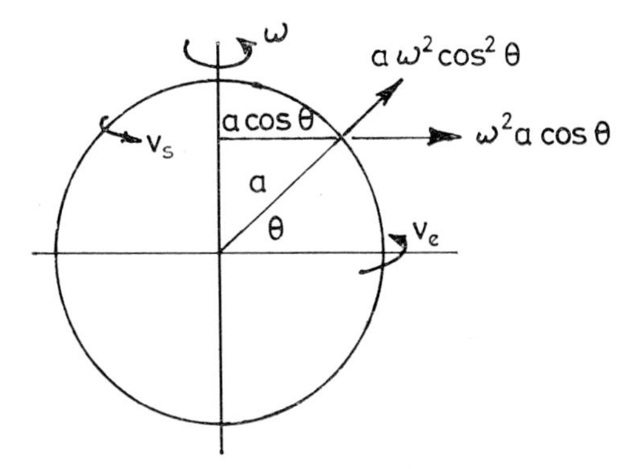

Fig. 16. Effect of rotational velocity on observed gravity.

as $v_e \cos^2 \theta / a$ where v_e is the speed of a point in the surface at the equator; v_e is about 450 ms^{-1}. If a ship moves over the surface with an east–west component of velocity v_s, the angular velocity of the ship about the polar axis is $[(v_e/a)+(v_s/r)]$ where $r = a \cos \theta$.

Thus the component of the centrifugal acceleration directed away from the centre of the Earth becomes :

$$a\left(\frac{v_e}{a}+\frac{v_s}{a\cos\theta}\right)^2\cos^2\theta,$$

which differs from the value for a stationary point by :

$$\frac{2v_e v_s}{a}\cos^2\theta.$$

The measured acceleration is increased if the ship moves from east to west, for the ship is moving in the opposite direction to the Earth, reducing the velocity and so decreasing the amount by which g is reduced by the centrifugal acceleration. The effect amounts to 7 mgal per knot at the equator. On account of ocean currents it is difficult to know the speed of a ship over the ground to better than a few knots in the deep oceans far from land, and it is clear that the correction for this speed, the Eötvös correction, is the major obstacle to the measurement of gravity at sea with anything approaching the accuracy achieved on land. An accuracy of 5 mgal in the deep oceans would be rather good.

For these reasons the problems of accurate navigation at sea is more important than that of the actual measurement of acceleration. The measurement problem is at the same time, more difficult than it is on land. There are three main issues—the stability of the instrument, the linearity of its response to accelerations and the alignment of the meter. The instrument must be more stable, that is the meter reading for a given value of gravity must be more nearly constant than for a land meter, because a ship-borne instrument will be away from a stable base for a very much longer time than a land instrument—a change of no more than a few milligals in many months is desirable.

With all springs the relation between extension and tension changes slightly with extension, mainly because the angle of the helix changes as the spring is stretched. This does not matter much in a land instrument where the changes of gravity are quite small and where the response of the spring can in any case be measured by comparison with known values of gravity. The case is quite different for a sea-borne meter where, on account of the motions of the ship, the range of accelerations is a large fraction of gravity. Suppose that the meter reading, R, instead of being proportional to gravity, is given by :

$$R = R_0 + ag + bg^2.$$

Suppose, also, that the vertical acceleration of the ship is :

$$g + A\cos\nu t$$

where A is the amplitude of the acceleration due to the motion of the ship and ν is the angular frequency of the motion. The instantaneous meter reading is then

$$R_i = R_0 + a(g + A\cos\nu t) + b(g^2 + 2gA\cos\nu t + A^2\cos\nu t),$$

and the value averaged over a number of periods of the motion of the ship is :

$$R_{av} = R_0 + ag + b(g^2 + \tfrac{1}{2}A^2).$$

Since A may be as much as one-tenth of g, it is important that the response of the meter should be very accurately linear.

Lastly, there is the effect of the horizontal motions of the ship, for the accelerations due to the actions of the waves are not solely vertical, in fact the horizontal accelerations have comparable amplitudes. Suppose that the horizontal acceleration at a meter is $\ddot{x}$ and that the vertical acceleration due to the wave motion is $\ddot{z}$; the total instantaneous acceleration at the meter is the resultant of $g + \ddot{z}$ vertically and $\ddot{x}$ horizontally (fig. 17). If the meter hangs freely in gimbals so that it

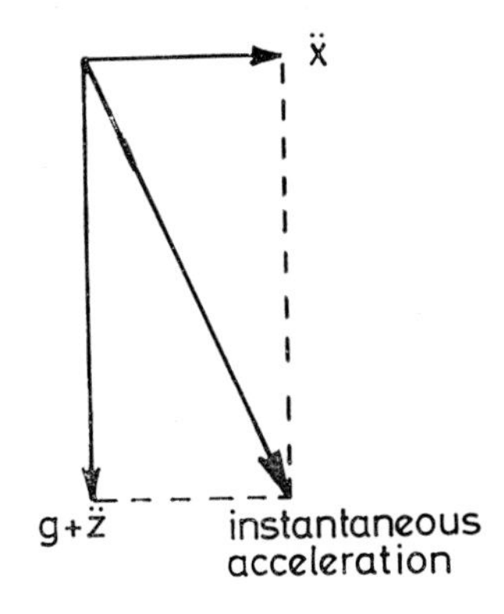

Fig. 17. Acceleration due to motions of a ship.

always indicates the resultant acceleration, its instantaneous reading will be proportional to :

$$(g^2 + 2g\ddot{z} + \ddot{z}^2 + \ddot{x}^2)^{\frac{1}{2}} \quad \text{or} \quad g\{1 + \ddot{z}/g + \tfrac{1}{2}(\ddot{z}^2 + \ddot{x}^2)/g^2 + \ldots\},$$

assuming that the meter responds linearly to acceleration. The average reading will be proportional to :

$$g + \tfrac{1}{2}\langle \ddot{z}^2 + \ddot{x}^2 \rangle_{av}/g,$$

from which it will be seen that separate measurement must be made of the ship accelerations in order to apply corrections for them. This can be done with accelerometers that respond to periodic accelerations. Alternatively, the meter may be mounted on a platform that is stabilized, usually by reference to a gyroscope, so that it remains normal to the direction of g, irrespective of the periodic accelerations of the ship. The response of the meter in that case is proportional to :

$$g + \ddot{z},$$

the mean value of which is g. Nowadays, very good stable platforms can be made and this is the preferred way of using a gravity meter.

Problems

2.1. In a symmetrical free-motion determination of gravity, the distance between the levels at which the motion of the moving object was timed was 1 m ; the time intervals between the upward and downward crossings were :

at the upper level 0·20400 s,
at the lower level 0·92572 s.

Calculate the value of gravity. (9·8118 m/s².)

2.2. The period of an invariable pendulum was 0·506065 s at a site where the value of gravity was 9·81265 m/s² and 0·505981 s at a second site. What is the difference of gravity between the two sites ? (325 mgal.)

2.3. A ship is steaming eastwards at 20 km/h at latitude 45°. By how much does the value of gravity measured on the ship differ from the value that would be measured on a stationary platform ? (−56 mgal.) Take the radius of the Earth to be 6400 km.

CHAPTER 3

gravity and the crust of the Earth

Gravity anomalies

THE Earth is approximately a sphere in which the density depends only on the radius so that its attraction is nearly the same as that of a point of the same mass as the Earth placed at the centre of the Earth; g is very nearly equal to GM/r^2, where r is the distance from the centre of the Earth. If gravity is measured at a height h above sea level, the value would be expected to be related to that at sea level as follows:

$$\frac{g_h}{g_0} = \frac{R^2}{(R+h)^2},$$

where g_0 is the value at sea level, g_h is that at height h, and R is the radius of the Earth. Then

$$g_h - g_0 = -\frac{2g_0 h}{R}.$$

To allow for the unavoidable fact that gravity measurements are made at different heights, they may be amended by adding to them the quantity:

$$\frac{2g_0 h}{R}.$$

The Earth is not however a sphere but is flattened at the poles, in consequence of which, as will be seen in Chapter 5, gravity increases from the equator to the poles by an amount which is proportional to $\sin^2 \phi$ where ϕ is the latitude. If we wish to compare values of gravity at different latitudes, we should apply a correction for this variation which we already know about and which, in the latitudes of Britain, amounts to about 1 mgal for every kilometre travelled northwards.

According to these simple ideas, the value of gravity at a height h and a latitude ϕ would be equal to:

$$g_e(1+\beta \sin^2 \phi) - 2g_e h/R,$$

where g_e is the value of gravity at the equator and β is a constant equal to about 0·0053.

If g_m is the measured value of gravity at some place, then the difference between g_m and the value expected on the simple ideas just set out, namely

$$g_m - g_e(1+\beta \sin^2 \phi) + 2g_e h/R,$$

is known as the *free air anomaly*, because it is the difference from the value to be expected if there were no attracting matter between the level of the sea and the point at which gravity is measured. In fact, of course, the attraction of the ground above sea level increases the value of gravity somewhat, by an amount which can often be calculated quite accurately. In flat country, that attraction is in fact $2\pi G\rho h$, where ρ is the density of the ground. In any case, let the attraction of the ground be called B. The expected value of gravity is then

$$g_e(1+\beta \sin^2 \phi)-2g_e h/R+B,$$

and the difference between the measured value and this calculated value is known as the *Bouguer anomaly*, after the French geodesist P. Bouguer who first studied the gravitational attraction of mountains.

Gravity and local geology

In country like the British Isles, it is found that the free air anomalies are generally greater in higher country, showing the effect of the attraction of the ground above sea level. The Bouguer anomalies, on the other hand, show little variation with height, but do vary with the densities of the rocks. When the attraction of the ground above sea level has been allowed for, the remaining variation of gravity, the Bouguer anomaly, shows the effects of the different densities of the rocks below sea level. Let us look at some simple instances. In the country between Leicester and Birmingham there are some small coalfields in which the Coal Measure rocks are slightly less dense than the older, harder rocks in the hollows of which they lie (fig. 18).

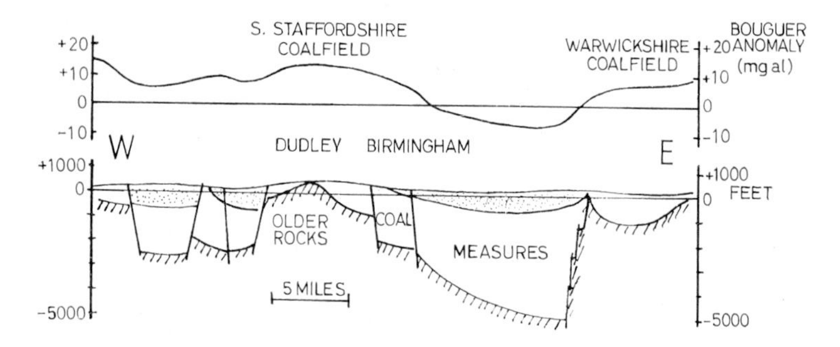

Fig. 18. Geology and Bouguer anomalies over coalfields of Warwickshire.

The surface dividing the Coal Measures from the older rocks slopes only gently and at any place above the Coal Measures, the value of gravity is reduced by the difference between the attraction of the Coal Measures and that of an equal thickness of the older rocks. The Bouguer anomalies are therefore less over the coalfields than they are over the older rocks between the fields, as shown in the diagram.

Because the thickness changes only slowly, we expect the reduction of the Bouguer anomaly at any point to be equal to the (negative) attraction of a slab of material having a thickness equal to that of the Coal Measures and a density equal to the difference between that of the Coal Measures and that of the older rocks.

The attraction of a slab of constant thickness and uniform density and having a lateral extent large compared with its thickness can be calculated straightforwardly. Let the thickness be t and the density ρ (fig. 19) and consider the attraction of a vertical cylindrical shell with its axis drawn through the point P at which the attraction is to

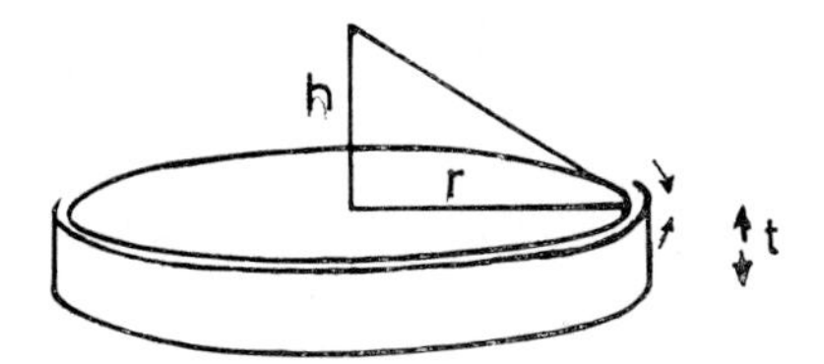

Fig. 19. Calculation of the attraction of an infinite slab of thickness t.

be found. Let that point be at a height h above the top of the slab and let the inner radius of the cylinder be r and the outer radius be $r+dr$. Consider a section of the cylinder that subtends an angle $d\theta$ at the centre. The volume of this element is $trd\theta dr$, its mass is $\rho trd\theta dr$, and the attraction it exerts at P is $G\rho trd\theta dr/(r^2+h^2)$ directed towards the element. The vertical component is :

$$\frac{G\rho tr\, d\theta\, dr}{r^2+h^2}\cos\phi,$$

where

$$\cos\phi = \frac{h}{(r^2+h^2)^{\frac{1}{2}}},$$

and the total vertical component is therefore :

$$G\rho t\int_0^{2\pi} d\theta\frac{hr\,dr}{(r^2+h^2)^{3/2}} = 2\pi G\rho th\frac{r\,dr}{(r^2+h^2)^{3/2}}.$$

To obtain the attraction of the whole slab, we integrate this expression over values of r from 0 to ∞ :

$$g = 2\pi G\rho th\int_0^{\infty}\frac{r\,dr}{(r^2+h^2)^{3/2}} = 2\pi G\rho th\left[\frac{1}{(r^2+h^2)^{\frac{1}{2}}}\right]_0^{\infty}.$$

The expression for the value of gravity close to the top of a simple slab is therefore :

$$2\pi G\rho t,$$

as given above.

If instead of being far from the edges of the slab, P is over a vertical edge, the attraction is just $\pi G\rho t$ (θ runs from 0 to π) and it rises towards the interior of the slab and falls off outside the slab more or less sharply according to how far P is above the top of the slab.

Such a variation of gravity is found when a fault forms an abrupt edge to a slab of rock. In general faults are not vertical and if they slope the variation of gravity is not symmetrical about the half value as it is for a vertical edge. A striking example is found at the Malvern Hills in Worcestershire (fig. 20). The rocks to the west are old hard dense rocks and those to the east are softer less dense sandstones.

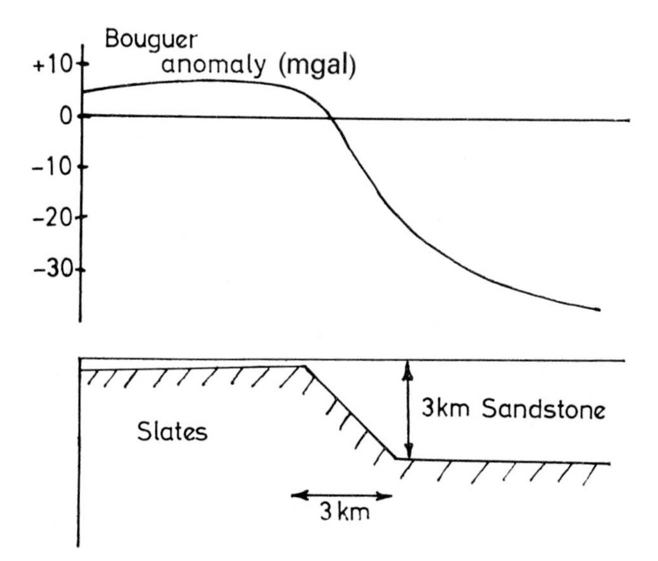

Fig. 20. Bouguer anomalies over the Malvern Hills.

The boundary between them slopes down to the east. The thickness of the sandstones is about 3000 m (10,000 ft) at the deepest. The rate at which gravity decreases to the east is greatest at the Malvern Hills, corresponding to the fact that the boundary between the two sorts of rocks is sloping down to the east.

In calculating the attraction of a slab of rock that slopes very gently in any direction, it is possible to neglect the slope and to suppose that the slab extends very far in all directions with the same thickness. When the slopes cannot be neglected, the next simplest situation is that in which the slab can be supposed to have a constant thickness with boundaries extending very far in either direction. On this basis, for example, the attraction of the rocks in the neighbourhood of the Malvern Hills can be calculated. Many bodies of rock cannot be taken to extend very far in any direction and the presence of nearby boundaries affects the attraction of gravity at all places above and in the neighbourhood of the mass of rock. While it is possible with a large computer to calculate the attraction of a body of rock of any size and shape, it is simplest to begin with to imagine the mass replaced by some symmetrical body for which calculations can be made algebraically and for which many tables and graphs are available to show the general nature of the attraction to be expected. A vertical cylinder or truncated cone will represent many geological structures such as bodies of granite or other igneous rocks. Granite bodies are particularly interesting for it is found that although they may seem to be made of very hard dense material, gravity over them is usually less, and frequently much less than over the surrounding rocks. In fact, the density of granite (about 2650 kg/m^3 or 2·65 g/cm^3) is slightly less by about 50 kg/m^3 (0·05 g/cm^3) than that of the rocks in which the granite is embedded, and because the granite extends to great depths, perhaps 20 to 30 km, gravity over it may be as much as 50 mgal less than over the surrounding rocks. In the British Isles, such low values of gravity are found over the granites of Dartmoor, of the Wicklow Mountains and of Donegal and of Scotland, while there are striking examples in the Rocky Mountains of North America. Other types of igneous rocks, especially those known as basic which contain much less silica and much more iron and magnesium than the granites, have higher densities than the surrounding rocks such as slates and sandstones, and in such cases, gravity is higher over the igneous rocks. There are examples in the Highlands of Scotland, and again in Ireland, where the dense rocks that lie under the extinct volcano of Slieve Gullion near the Mourne Mountains, are shown by the very high values of gravity in the neighbourhood (fig. 21).

It can be seen from these examples that when two rocks differ in density, the values of gravity over them will vary according to the geological structure of the rocks, and so may be used to help work out the structure where it cannot be seen from the surface; in particular, gravity measurements are often used in prospecting for oil and minerals. Gravity surveys have been widely used in searching for oil bearing rocks although usually they only give a first indication that suitable structures are present in which oil may be trapped; more detailed investigations must then be made by seismic methods and by drilling boreholes. Most of the areas with exposed or concealed coalfields in Britain have thus been surveyed since the last war and the results have also helped to indicate extensions of the coalfields. One type of structure that often shows up clearly is the salt dome. They are found for example in the Gulf Coast of North America and in the plain of North Germany, where beds of rock salt lying between sandstones or clays have been squeezed up into round plugs in the overlying rocks (fig. 22). The domes, which often trap oil or gas, are revealed by low values of gravity because the

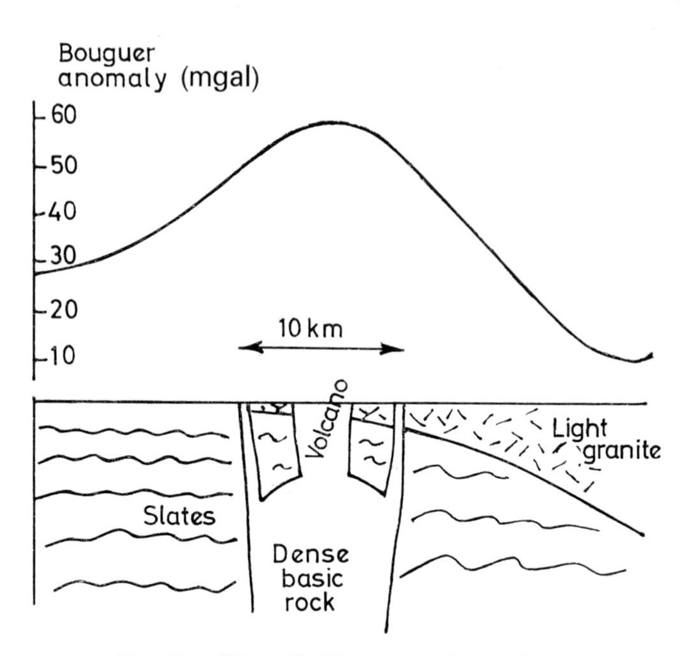

Fig. 21. Slieve Gullion—an ancient volcano.

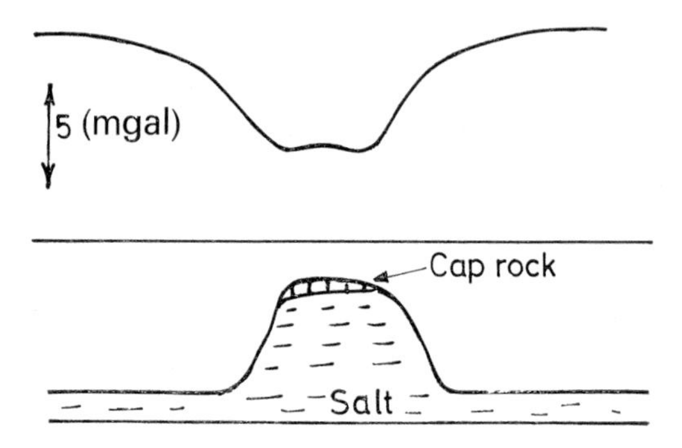

Fig. 22. Gravity anomalies over a salt dome.

salt is less dense than the sandstones or other rocks. Gravity surveys may also help to indicate the presence of minerals, especially of dense metallic ore deposits, and have been much used for this purpose in Sweden.

Gravity measurements by themselves can only show the presence of rocks of different densities and may be able to give some indication of their forms and thicknesses but without other information it is not possible to say what sort of rocks they are geologically. The extra information needed to interpret the gravity survey geologically may be just a general knowledge of the geology of the area or it may come from other geophysical surveys, but in any case, geophysical investigations are most effective when different methods are combined. Interpretations of gravity surveys are subject to a quite general limitation in that any variation of gravity observed at the surface can be due to a whole range of possible distributions of rocks of different density at depth, in particular, a shallow structure with a large difference of density can always be found that gives the same variation of gravity as a deep structure with small differences of density. The converse is not true, and the rate at which gravity changes limits the maximum depth at which the rocks that cause it can lie. This apart, variations of gravity can be interpreted in a wide variety of ways and other information, for example values of the densities of different rocks, must be used to limit the variety.

Gravity in mountain ranges

The structures of which we have been thinking so far are for the most part not more than a hundred kilometres in the least horizontal extent and are often much smaller. When we come to look at the variation of gravity over larger areas of the surface of the Earth, we find that it is no longer possible to understand it in the same way. We have seen that on the small scale, the effect of the attraction of the hills and valleys can be removed by calculating the Bouguer anomalies and that those anomalies can then be interpreted as due to the variations of the densities of rocks within usually 20 km of the surface. The Bouguer anomalies will usually be found to be less than 50 mgal. If we look at values of gravity in the great mountain ranges such as the Alps, the Rocky Mountains or the Himalayas, we find a quite different situation. The Bouguer anomalies now vary with the height of the mountains, being least where the mountains are highest, the variation being such that the free air anomalies are almost constant. In calculating the Bouguer anomalies, it seems that too great an allowance has been made for the attraction of the rocks of the mountains and in fact, the values of gravity are little different from what they would be if the mountains were not there at all (fig. 23). These remarkable observations were first suggested long before measurements of differences of gravity could be made with any ease or accuracy and the name, Bouguer anomaly, recalls the early days of geophysics. The French savant, Bouguer, had suggested that the mean density of the Earth could be found by comparing its attraction with that of a mountain of which the attraction could be calculated and had tried out this idea on Mt. Chimborazo in the Andes (1749). Subsequently Maskelyne, a British scientist (1774), studied the attraction of the mountain Schiehallion in Scotland (fig. 24). If a simple pendulum

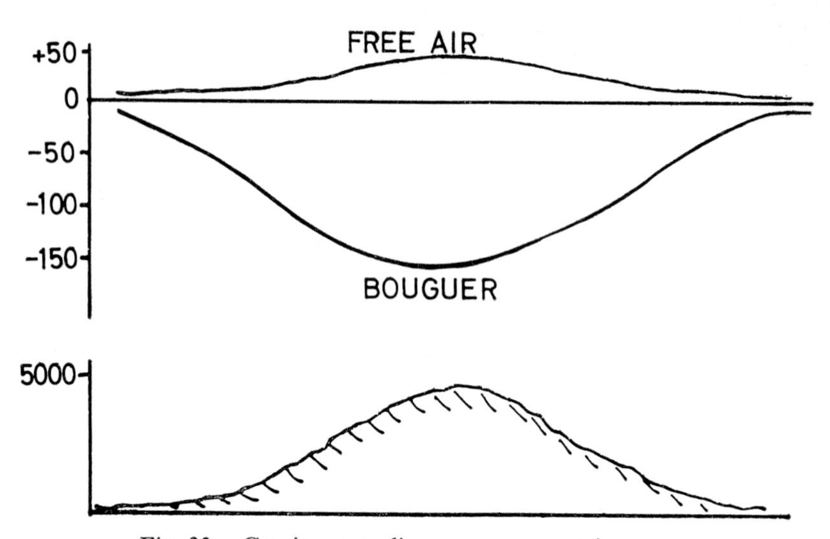

Fig. 23. Gravity anomalies across a mountain range.

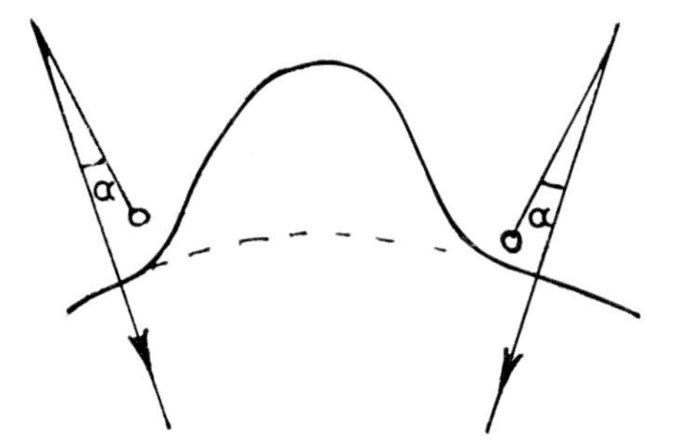

Fig. 24. Horizontal attraction of a mountain on a plumb bob.

or plumb bob is hung up near the mountain, it will be pulled to one side through an angle α given by :

$$\alpha = f_M/g,$$

where g is the attraction of the Earth, that is the value of gravity at the point, and f_M is the attraction of the mountain horizontally. The undisturbed direction in which the plumb bob would hang is not known, but if it is moved to the other side of the mountain, it will be pulled aside through an angle $-\alpha$. Suppose that the directions of the plumb bob in the two positions are found relative to the direction of some fixed star. Also let the distance apart of the two positions of the plumb bob along the surface of the Earth be l. Since the undisturbed plumb bob would point to the centre of the Earth (assumed to be spherical), the change in its direction relative to the fixed stars should be l/R, where R is the radius of the Earth. The actual observed change would be $(l/R)+2\alpha$; thus α and therefore f_M/g could be found. Schiehallion is a small and relatively isolated mountain and Maskelyne was able to survey it, to calculate its horizontal attraction on the plumb bob and to derive a not unreasonable value for the mass of the Earth. When the same experiment was tried by the French scientist Petit in the Pyrenees (1849) he found that the plumb bobs, instead of being drawn towards the mountain, were apparently repelled by it. Subsequently, Archdeacon Pratt (1855) found that surveys of the Himalayas gave similar results. These studies, confirmed subsequently by Hayford in the U.S.A. (1909), showed that mountains behaved as if the density of the material above sea level were very much less than that of normal rocks, or even negative. Pratt suggested that below mountains the density of the material of the Earth was less than elsewhere (an idea advanced by Cavendish in 1773), so that the total mass in any vertical column was the same whether there were mountains above sea level or not. If that is the case, the value of gravity in the neighbourhood of mountains will show little effect of the attraction of the mountains themselves. The situation is very like that of a solid floating in a liquid—if it is of irregular shape, it adjusts its attitude so that the total mass of solid and liquid in any vertical column is the same and if gravity were measured above such a floating solid, it would be found scarcely to differ whether it were measured above the free surface of the liquid or anywhere above the solid. The principle that the mass of a mountain is compensated by a defect of mass at some depth in the Earth is known as the principle of *Isostasy* and as a result of combining the results of gravity measurements with those of seismic studies, we now have a rather clear idea of the structure of mountain ranges and how they come to be in isostatic balance.

The deep structure of the Earth

Sound waves travel at different velocities through different rocks, in general going faster the harder the rocks, and by studying the times taken for the sound waves from earthquakes or artificial explosions to reach different points on the surface of the Earth, the positions of the boundaries between rocks having different velocities can be worked out. It is found that at a depth of about 30 km below the surface of continents there is a very well-defined boundary that separates rocks similar to those seen at the surface, the crust of the Earth, from deeper rocks in which sound travels at a much higher speed. The layers between this discontinuity and the fluid core of the Earth are known as the *mantle*, those in the upper part to a depth of about 1400 km being the *upper mantle*. These major divisions of the Earth into the fluid core (with a solid inner core at the centre), lower mantle, upper mantle and crust, are well established as a result of studies of the times taken by sound waves to travel through the Earth (fig. 25).

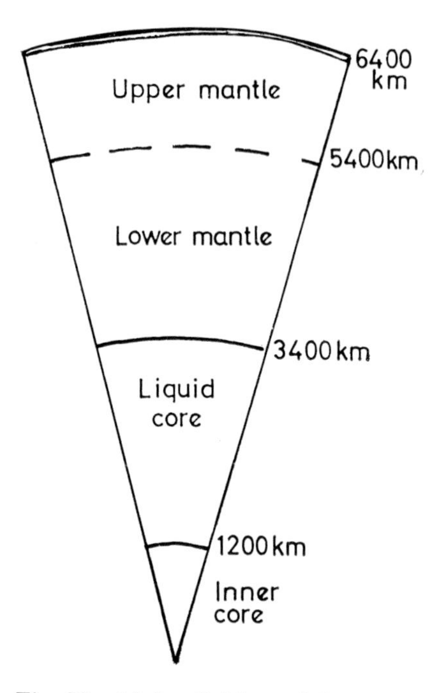

Fig. 25. Major divisions of the Earth.

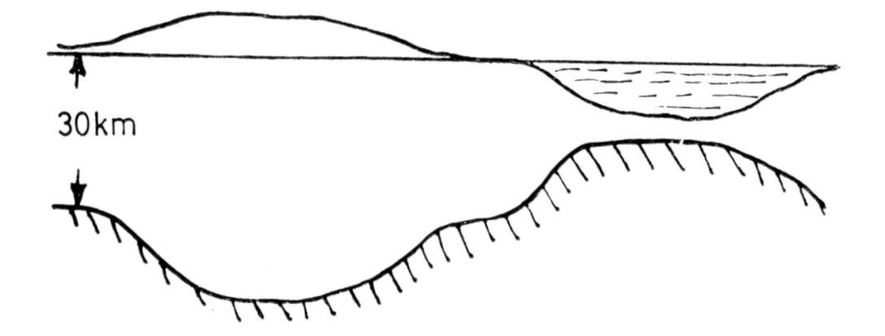

Fig. 26. Principle of isostatic balance.

Isostasy

Isostatic balance comes about because under mountain ranges the boundary between the crust and the upper mantle lies at a greater depth than elsewhere. The rocks of the upper mantle are denser than those of the crust and therefore there is a deficit of mass below the mountains that compensates for the mass in the mountains themselves (fig. 26). Mountains are places where the crust of the Earth is thicker than normal in the continents, part of the extra thickness standing up above the surface as the mountains, but most (as with an iceberg) projecting down into the upper mantle.

Gravity measurements at sea show that the same principle of isostasy applies as between the continents and the oceans generally. Free air gravity anomalies are little different whether they are measured over the continents or over the oceans, despite the fact that mass of the oceans is less than that of the continents by an amount corresponding to a slab of thickness equal to the depths of the oceans (say 5 km) and of density equal to the difference between the density of the crust, about 2650 kg/m^3, and that of sea water, slightly greater than 1000 kg/m^3. This is a mass far greater than that of any mountain range. To compensate for the apparent deficit of mass and yet to maintain the attraction of gravity unchanged, there must be some extra mass below the oceans. With the help again of seismic studies, the extra mass is found to be accommodated by a reduction of the thickness of the crust under the oceans, or even its complete absence, so that the denser materials of the upper mantle replace those of the crust under the oceans (fig. 27). Continents and oceans are thus seen to be quite different, and to explain how these large and somewhat irregular divisions of the surface of the Earth came into existence is one of the major problems of geophysics.

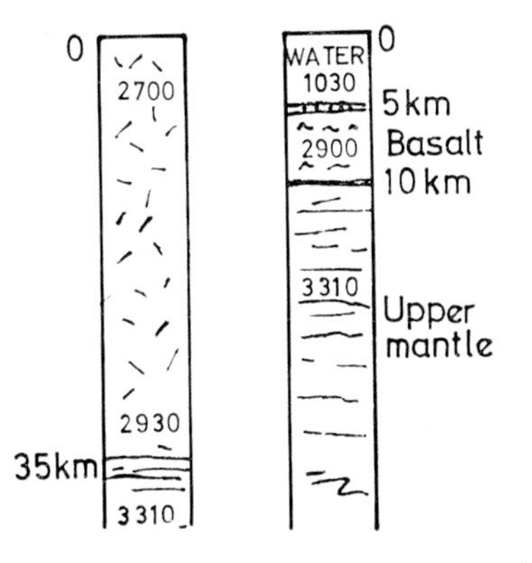

Fig. 27. Structures of the oceans and continents. (The figures 2700, etc., are densities in kg/m^3).

Deformations of the Earth's crust

The crust of the Earth appears to behave quite differently according to whether we look at its small scale or its large scale features. When we study local geology, the crust appears to support any extra load that may be put on it and we see the full effect of that load in the changes of the attraction of gravity over it, but when we look at the larger areas of the crust, then it seems that the crust cannot support any extra load and it sinks into the mantle as into a liquid until the upward hydrostatic pressure of the denser mantle balances the extra load on the crust. This behaviour is not in fact very surprising for it means that the crust of the Earth, like any other solid plate, can support loads up to a certain amount after which it will yield. The loads applied by local geological structures are less than the yield strength of the crust and those applied by mountain ranges and by the continents and oceans are much greater. Examples of loads comparable with the yield strength of the crust can be found, especially in the

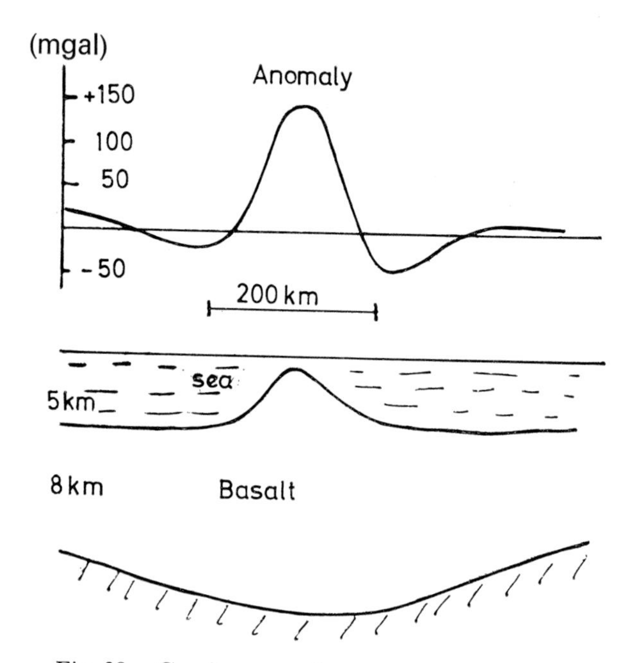

Fig. 28. Gravity anomalies over an oceanic island.

oceans. Gravity measurements around isolated islands show that the load of the island is partly but not completely compensated and that the compensation extends to some distance away from the island, corresponding to the fact that the thin crust of the ocean is pushed into the mantle material like an elastic plate (fig. 28). Another group of structures cannot be maintained by isostatic balance, even allowing for some support from the elastic bending of the crust. In the West Indies and in the East Indies, groups of islands are found to lie in curved chains (island arcs) in association with volcanoes. Many earthquakes occur in the areas. Parallel to the chains of islands there are furrows in the sea floor where the oceans are much deeper than the average (fig. 29). The attraction of gravity is very low over the ocean trenches which are not only not compensated isostatically, but seem to have even lighter rocks below them. It is thought that light material such as sand and mud is accumulating on the sea floor in the island arcs and is gradually being forced down into the mantle as a result of horizontal forces in the crust driving it in towards the

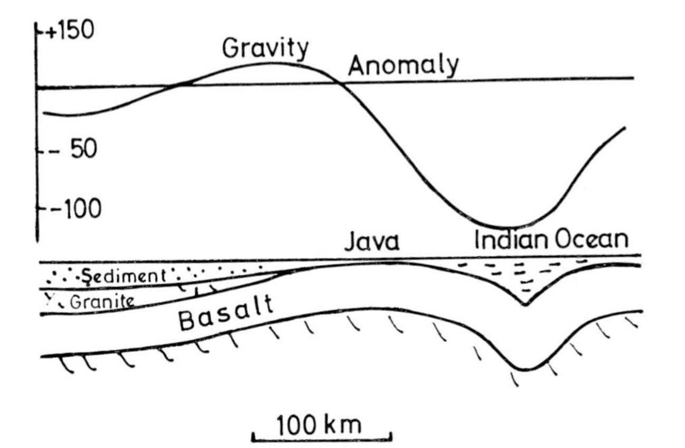

Fig. 29. Section across an island arc (East Indies).

trenches from both sides. The boundaries between the continents and oceans show somewhat similar features, gravity over the continental margins being rather low and corresponding to accumulations of mud and sand washed down from the land and deposited at the edge of the continental shelf without being fully compensated by extra mass in the mantle. In course of time the sediments in the island arcs and to a less extent those at the continental margins will become hardened by heat and pressure and, as a result of the forces in the crust, will be forced up above the seas to form mountain ranges like the Alps or the Rockies.

The world-wide variation of gravity

So far we have tacitly assumed that in calculating anomalies we know exactly how gravity varies with latitude due to the flattening of the poles and the rotation of the Earth. In studying local geology, defects in our knowledge of that variation will not really affect our study of the geology but when we come to the differences between continents and oceans, it starts to become difficult to separate the world-wide variation of gravity from the effects of the structures we wish to examine. The best we can do is to take a world-wide average of the variation of gravity with latitude and to hope that the effects of continents and oceans will indeed average out. For this purpose we should like to have values of gravity that cover the Earth with a uniform density, so many observations in each equal area of the surface wherever that area might be, whether on land or on sea. However, measurements at sea are much more difficult to make than those on land, they are much less accurate because of the problems of navigation and of the correction for the speed of the ship, and the seas cover about three-quarters of the area of the Earth. Because measurements on land are much more numerous and more accurate than those at sea, any world-wide average behaviour of gravity that we derive from measurements at the surface of the Earth, will probably not be a true average treating the seas and oceans equally, but will most likely apply more to the continents than to the seas. Gravity measurements made at the surface may therefore give a slightly misleading idea of the differences between continents and oceans and if we wish to go farther and ask what do the residual variations of gravity tell us about the mantle of the Earth after we have allowed for what we know of the structure of the crust, then it will be most difficult to learn anything significant from surface gravity measurements. Artificial satellites have given us the opportunity to study the world-wide variation of gravity through observations that are not restricted to selected parts of the surface of the Earth.

Problems

3.1. By how much does gravity decrease in going up a mountain 1000 m high? Take g to be $9{\cdot}8$ m/s^2 and the radius of the Earth to be 6400 km. (306 mgal.)

3.2. By how much does gravity increase in going from a point at sea level in latitude 50° to one at sea level in latitude 55°? Take the factor β to be $5{\cdot}3 \times 10^{-3}$ and g_e to be $9{\cdot}78$ m/s^2. (410 mgal.)

3.3. Coal Measures 2000 m thick and of density 2500 kg/m^3 are surrounded by older rocks of density 2700 kg/m^3. Neglecting the effects of the edges of the coal field, by how much is gravity less over the coal measures than over the older rocks? (17·4 mgal.)

$$G = 6{\cdot}67 \times 10^{-11} \text{ Nm}^2/\text{kg}^2 *.$$

3.4. In oceanic areas the upper 50 km of the Earth consists of:
sea water, density 1000 kg/m^3, thickness 5 km,
sediments, density 2500 kg/m^3, thickness 2 km,
oceanic crust, density 3000 kg/m^3, thickness 5 km,
upper mantle, density 3300 kg/m^3, thickness remainder.

In continental areas, the same thickness is made up of:
continental crust, density 2700 kg/m^3, thickness T km,
upper mantle, density 3300 kg/m^3, thickness $50-T$ km.

Find T for isostatic equilibrium—equal mass above the level of 50 km. (24·3 km.)

---

* 1 Newton (N) is 1 kg m/s^2 and so 1 Nm^2/kg^2 = 1 m^3/kg s^2. (p. 2.)

CHAPTER 4

gravity and satellites

Elliptical orbits of satellites

ONE of the first uses made of artificial satellites was to derive from their behaviour information about gravity just outside the Earth. As has been emphasized in earlier chapters, the Earth is very nearly spherically symmetrical and gravity outside it is very nearly the same as if the whole mass of the Earth were concentrated in a point at the centre ; if this were strictly so, then satellites would move around the Earth in elliptical paths according to Kepler's laws. Because most satellites are relatively close to the Earth, they are much more sensitive to small irregularities in the gravity field than is the natural satellite, the Moon, and being close to the Earth, the small departures of the path from an ellipse that correspond to the irregularities of gravity, are many of them easy to observe.

Let us first see how Kepler's laws are derived. In the following mathematical working it will be shown that a satellite moving about a point mass moves in an elliptical orbit, and the formula relating the period and size of the orbit to the mass of the attracting body will be derived. The way in which the orbit is placed in space is taken up in the next section.

Suppose that the mass of the Earth is indeed all concentrated at the centre. Let it be M and let a satellite of mass m be at a distance r. (fig. 30) Let the direction Mm make an angle θ with some fixed direction. m moves in a plane because the acceleration of m is directed along the line mM and there is no acceleration perpendicular to that line ; m will continue to move in the plane defined by Mm and its initial velocity.

The velocities of m are $\dot{r}$ in the direction Mm and $r\dot{\theta}$ in the perpendicular direction. (The dot denotes differentiation with respect to time.) Consider now the acceleration of m. Firstly, there is the rate of change of $\dot{r}$ in the direction of increasing r, namely $\ddot{r}$. Then because of the rotation of r at the rate $\dot{\theta}$, the direction of the velocity component $r\dot{\theta}$ is also changing at the same rate, a change that has a component $r\dot{\theta}\times\dot{\theta}$ in the direction of decreasing r. The total acceleration in that direction is therefore :

$$\ddot{r}-r\dot{\theta}^2.$$

Similarly, the rate of change of the velocity $r\dot{\theta}$ in the direction perpendicular to r is $(d/dt)(r\dot{\theta})$, that is $\dot{r}\dot{\theta}+r\ddot{\theta}$, while the component in the

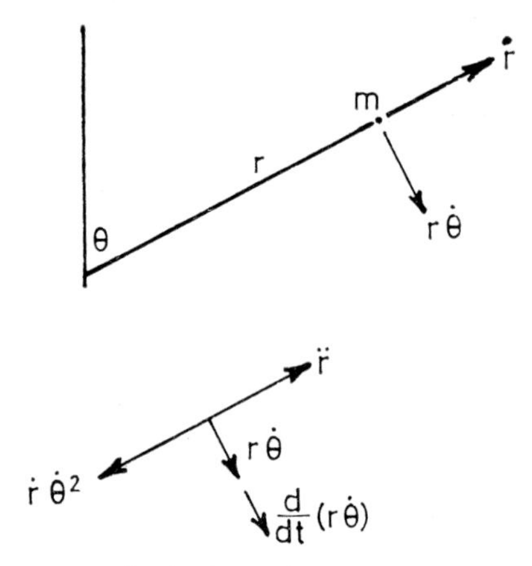

Fig. 30. Components of velocity and acceleration in polar co-ordinates.

same direction arising from the rotation of the velocity $\dot{r}$ is $\dot{r}\dot{\theta}$. Thus the total acceleration perpendicular to r is :

$$2\dot{r}\dot{\theta}+r\ddot{\theta}.$$

Since there is no force perpendicular to r :

$$2\dot{r}\dot{\theta}+r\ddot{\theta} = 0.$$

that is :

$$2\frac{\dot{r}}{r}+\frac{\ddot{\theta}}{\dot{\theta}} = 0.$$

Integrating :

$$2 \ln r+\ln \dot{\theta} = \text{a const.},$$

so that

$$\ln (r^2\dot{\theta}) = \text{a const.},$$

or

$$r^2\dot{\theta} = h,$$

where h is a constant.

Now the angular momentum of m about M is :

$$m\times r\times (\text{perpendicular velocity}) = mr\times r\dot{\theta} \quad \text{or} \quad mr^2\dot{\theta},$$

and so h is the angular momentum of the satellite per unit mass. The angular momentum is a vector perpendicular to the plane in which the satellite moves and this result shows that it is constant.

Consider now the acceleration along r. It is directed towards M, that is, in the direction of decreasing r, and is equal to $-GM/r^2$. Putting GM equal to μ :

$$\ddot{r}-r\dot{\theta}^2 = -\mu/r^2.$$

But $\dot{\theta}^2 = h^2/r^4$ and so

$$\ddot{r}-\frac{h^2}{r^3} = -\frac{\mu}{r^2}.$$

This equation will give r as a function of the time, but if we are to find the shape of the path of the satellite, we need to find r as a function of the angle θ. Now we again make use of the result that $\dot{\theta} = h/r^2$ which tells us that

$$\frac{d}{dt} = \frac{h}{r^2} \cdot \frac{d}{d\theta}.$$

Then

$$\ddot{r} = \frac{d}{dt}\left(\frac{dr}{dt}\right) = \frac{h}{r^2}\frac{d}{d\theta}\left(\frac{h}{r^2}\frac{dr}{d\theta}\right).$$

For convenience, put $dr/d\theta = r'$ so that

$$\ddot{r} = \frac{h}{r^2}\frac{d}{d\theta}\left(\frac{h}{r^2}r'\right).$$

It turns out that it is easier to integrate the differential equation for r if instead of dealing with r, we change to $1/r$, which we shall call ρ.

Then

$$r' = \frac{d}{d\theta}\left(\frac{1}{\rho}\right) = -\frac{1}{\rho^2}\rho'$$

and

$$\ddot{r} = \frac{h}{r^2}\frac{d}{d\theta}(-h\rho') = -h^2\rho^2\rho''.$$

The equation

$$\ddot{r} - h^2/r^3 = -\mu/r^2$$

then becomes

$$-h^2\rho^2\rho'' - h^2\rho^3 = -\mu\rho^2,$$

that is :

$$\rho'' + \rho = \mu/h^2,$$

of which the solution is :

$$\rho = \frac{\mu}{h^2} + C\cos\theta,$$

where C is a constant ; if we put :

$$\frac{h}{\mu^2} = r_0,$$

then

$$r = \frac{r_0}{1 + e\cos\theta},$$

where

$$e = C/r_0.$$

This is the polar equation of an ellipse. In fig. 31, let F be the focus of the ellipse and let DD' be the directrix. Let P be any point on the ellipse and let PD' be perpendicular to the directrix. The definition of the ellipse is that

$$FP = ePD',$$

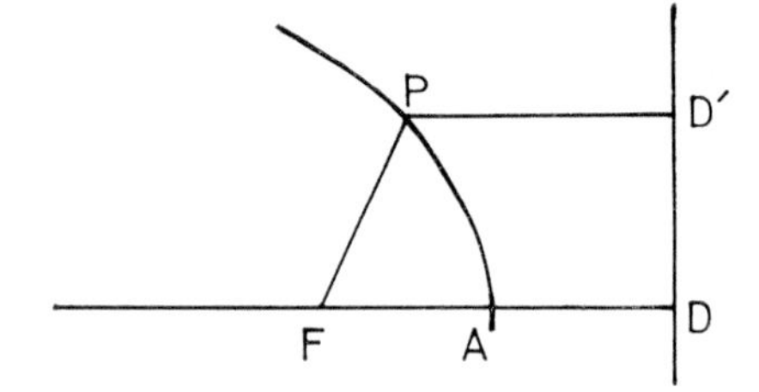

Fig. 31. Geometry of the ellipse in polar co-ordinates.

where e is the eccentricity. In particular, if FD is perpendicular to the directrix, and if it intersects the ellipse in A, then $FA = eAD$. Now if the angle θ is measured from FA, then

$$\begin{aligned} PF &= ePD' \\ &= e(FD - PF\cos\theta), \end{aligned}$$

or, putting PF equal to r:

$$r(1+e\cos\theta) = eFD,$$

which is the same as the result just derived for the orbit if eFD is identified with r_0.

Now let us calculate the lengths of the major and minor axes of the ellipse.

The major axis, $2a$, is found by adding together the values that r takes when θ has the values 0 and π, that is:

$$2a = r_0\left(\frac{1}{1+e}+\frac{1}{1-e}\right) = \frac{2r_0}{1-e^2}.$$

Let X be the foot of the perpendicular dropped from P to the major axis. The semi-major axis of the ellipse is identical with PX when the latter has its maximum value. Now

$$PX = r_0 \sin\theta/(1+e\cos\theta).$$

By differentiating this expression with respect to θ, you will find that PX reaches a maximum when $\cos\theta = -e$, and so b, the semi-minor axis, is equal to:

$$\frac{r_0}{(1-e^2)^{\frac{1}{2}}}$$

If we choose the semi-major axis, a, and the eccentricity, e, of the ellipse, as the constants to describe its shape, its polar equation is:

$$r = \frac{a(1-e^2)}{1+e\cos\theta}.$$

The period of elliptical motion

We must now relate the size of the ellipse to the attraction of the central body, assumed to be much more massive than the satellite.

At any value of the radius r, the angular velocity $\dot{\theta}$, is given by the equation for angular momentum, namely :

$$\dot{\theta} = \frac{h}{r^2} = \frac{h}{r_0^2}(1+e\cos\theta)^2.$$

But $r_0 = h^2/\mu$ and therefore

$$\dot{\theta} = \frac{\mu^{1/2}}{r_0{}^{3/2}}(1+e\cos\theta)^2$$

$$= \frac{\mu^{1/2}}{a^{3/2}}\frac{(1+e\cos\theta)^2}{(1-e^2)^{3/2}},$$

since

$$r_0 = a(1-e^2).$$

Now the time taken by the satellite to describe one revolution in the orbit is given by :

$$\int_{\text{one revoln.}} dt = \int_0^{2\pi}\frac{dt}{d\theta}d\theta,$$

the integral extending from 0 to 2π because one complete revolution corresponds to an increase of θ by 2π.

Denoting the period for one revolution by T,

$$T = \frac{a^{3/2}}{\mu^{1/2}}(1-e^2)^{3/2}\int_0^{2\pi}\frac{d\theta}{(1+e\cos\theta)^2}$$

$$= \frac{a^{3/2}}{\mu^{1/2}}(1-e^2)^{3/2}\int_0^{2\pi}d\theta(1-2e\cos\theta+3e^2\cos^2\theta+\ldots)$$

$$= 2\pi\frac{a^{3/2}}{\mu^{1/2}}(1-e)^{3/2}(1+\tfrac{3}{2}e^2+\ldots),$$

or

$$T = 2\pi a^{3/2}/\mu^{1/2}.$$

This result is Kepler's third law of planetary motion and relates the size of the orbit (a) to the period (T) and the central mass (μ).

We now have a description of the orbit so far as its shape within its plane and the relation of its size to the mass of the central body. We may relate these quantities to the value of gravity at the surface of the central body, supposed to be the Earth. If we suppose the Earth to have spherical symmetry and if the radius of the surface is R, the value of gravity at the surface, g, is GM/R^2, and so $\mu = gR^2$.
Hence

$$T^2 = \frac{4\pi^2 a^3}{gR^2}.$$

These results enable us, as will be seen in Chapter 5, to use observations of artificial satellites, to derive either the size or the mass of the Earth.

The orbit in space

We must now place the orbit in space. To do so, we relate it to a plane which is fixed in space, or which moves only very slowly, and the plane we choose for artificial satellites of the Earth is the equator-plane of the Earth. This plane intersects the plane of the orbit of the Earth about the Sun, the *ecliptic plane*, in a line which points in a direction known as the First Point of Aries, a direction lying in the constellation Aries, and which rotates slowly relative to the fixed stars as a result of a slow motion of the axis of rotation of the Earth; the motion is called the precession of the equinoxes (fig. 32). Now

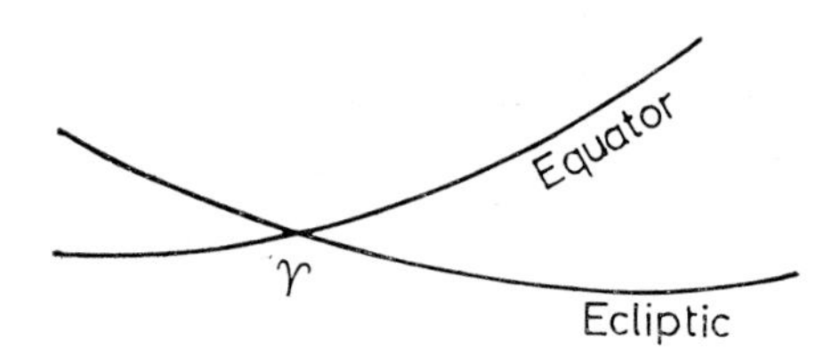

Fig. 32. The planes of the equator and ecliptic.

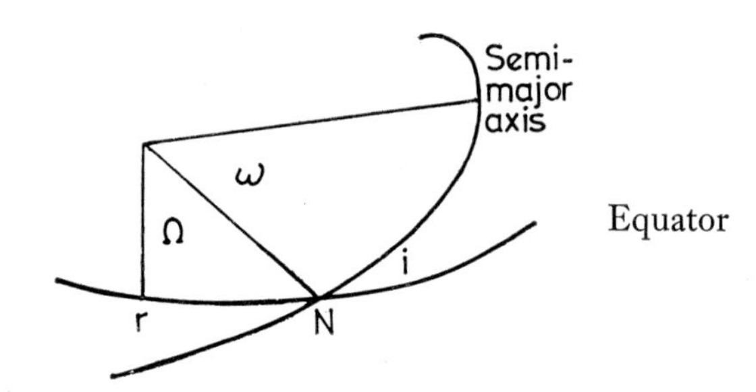

Fig. 33. The node, *N*, and inclination, *i*, of an orbit.

imagine the plane of the orbit of a satellite and the line, known as the *line of nodes*, in which it intersects the plane of the equator of the Earth (fig. 33). The angle between these two planes is called the inclination of the orbit of the satellite and is denoted by i. The angle, measured in the plane of the equator, between the line of nodes and the direction of the first point of Aries from the centre of the Earth is called the longitude of the node and is denoted by ☊. To be definite, the angle is measured to the ascending node, *N*, that is the direction of the satellite at the point at which it crosses the equator when going from south to north. The opposite direction is called the descending node.

Lastly, we have to define the direction in which the semi-major axis of the orbit lies. This is given by the angle ω measured in the plane of the orbit between the line of nodes and the direction of the point at which the satellite is closest to the focus, called *Perigee* for satellites about the Earth.

Orbits in the gravity field of the Earth

If the Earth were a spherically symmetrical body so that its attraction was the same as that of an equal mass at the centre, the orbit of any satellite would be an ellipse of constant size and orientation in space, and the satellite would go round it in a constant time, about 100 min for a satellite close to the Earth. Even observations that are quite easy to make from one's own back garden show that satellites do not behave in this constant way.

As the Earth spins on its axis, satellites are seen in different directions relative to the Sun from a point spinning with the Earth. Many satellites can therefore be seen lit up by the Sun at local dawn or dusk while the observer is still hidden from the Sun and so sees the stars at the same time as he sees the satellite (fig. 34). The observer can

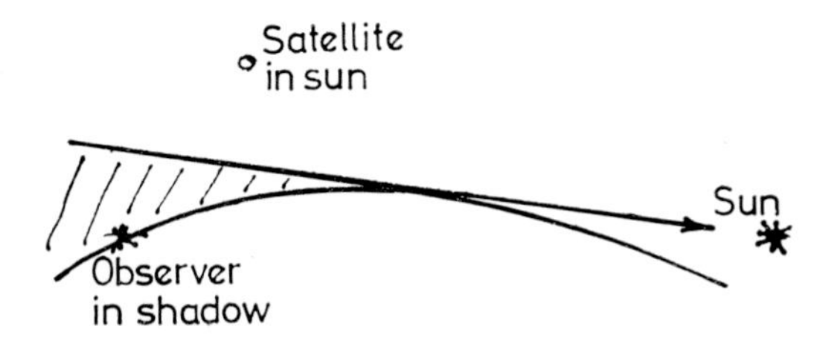

Fig. 34. A satellite in sunlight seen by an observer in shadow.

then record the changing direction of the satellite as time passes by noting its direction against the fixed stars. Since a number of directions define a plane, he can calculate the plane of the orbit and where it cuts the plane of the equator. By noting the times of the satellite on successive orbits, the observer can calculate the period. Such observations and calculations neglect finer points but even so they show quite clearly that satellites do not behave in the simple way predicted for a spherically symmetrical Earth. The period, T, gets steadily less, so that the semi-major axis, a, also decreases, and the line of nodes rotates steadily along the equator in the opposite direction to the motion of the satellite, at a rate of about 3° per day for a close satellite with an inclination of about 60°, one such as might be readily seen in the British Isles. The reduction of the period is due to the drag of the air on the satellite which continually slows it down and finally causes it to come in to the dense atmosphere where the heating due to the friction of the air causes the satellite to burn up.

Although this behaviour is very interesting and has told us a great deal about the very high atmosphere of the Earth, it does not have much influence on what we can learn about the gravity field of the Earth from the orbits of satellites and no more will be said about it.

The rotation of the plane of the orbit is, on the other hand, directly related to ways in which gravity differs from that on a spherically symmetrical Earth and is scarcely affected by the resistance of the air.

Satellites can also be observed by means of transmissions from any radio transmitters that they may carry. Such observations can be made continuously since they do not depend on the satellite being illuminated by the Sun while the observer is in shadow, and so give more detailed information about the satellite. Again the easiest observations are of successive directions of the satellite, using a radio interferometer direction finder; they show that the direction of the major axis also changes steadily, while, like the value of a, the value of the eccentricity e, also decreases steadily due to air drag. In addition, the inclination oscillates slightly about a fixed value.

Gravity about an Earth flattened at the poles

The steady changes of the orbit due to gravitational effects are in the main the result of the flattening of the Earth at the poles. In such a flattened Earth the moment of inertia about the axis of rotation is greater than that about any axis in the equator. Let us now calculate the potential of gravity of a model Earth which is symmetrical about the axis of revolution so that the moments of inertia about all axes in the plane of the equator are the same and the moment about the axis of rotation is somewhat greater. The mathematics may be omitted ; the result will be found at the end of this section.

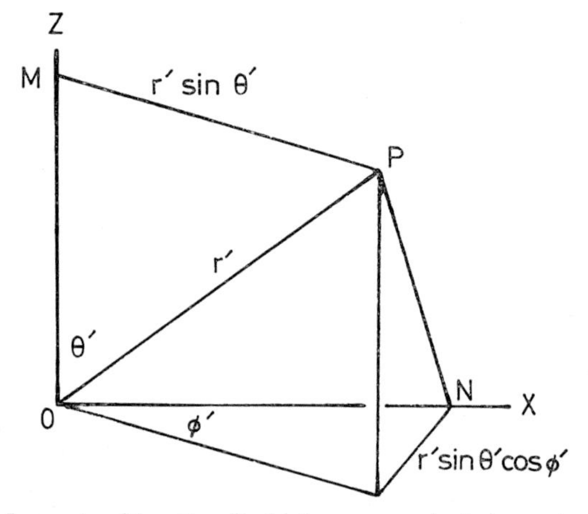

Fig. 35. Moments of inertia of a body symmetrical about the polar axis.

Let OZ (fig. 35) be the axis of rotation and let OX be any direction fixed in the equatorial plane. Let the position of any point P' be given by spherical polar co-ordinates, r' measured from the centre of mass O, θ' measured from the polar axis OZ, and the azimuth ϕ' measured from the direction OX.

The moment of inertia, C, about the polar axis is given by the integral of the product of the mass of an element of volume dv by the square of its distance $P'M$ from the polar axis. If σ is the density at the point with the co-ordinates r', θ', ϕ',

$$C = \int_V dv\, \sigma r'^2 \sin^2 \theta',$$

for $dv\,\sigma$ is the mass of the element of volume dv and $r' \sin \theta'$ is the perpendicular distance of the element from the polar axis. The integral is taken over the whole volume V.

Now $dv = r'^2 \sin \theta'\, dr'\, d\theta'\, d\phi'$ and so

$$C = \int dr' \int_0^\pi d\theta' \int_0^{2\pi} d\phi'\, r'^2 \sin \theta'\, \sigma r'^2 \sin^2 \theta'.$$

But because the body is symmetrical about the polar axis, the density σ does not depend on ϕ' and so we may integrate with respect to ϕ' without knowing anything more about σ. Since

$$\int_0^{2\pi} d\phi' = 2\pi,$$

$$C = 2\pi \int dr' \int_{-1}^{+1} d(\cos \theta') \sigma r'^4 \sin^2 \theta'.$$

The moment of inertia A about the axis OX in the equatorial plane is similarly equal to the integral of an element of mass multiplied by the square of the distance $P'N$ from the axis, that is by :

$$r'^2(\cos^2 \theta' + \sin^2 \theta' \cos^2 \phi').$$

Thus

$$A = \int_V dv \sigma r'^2(\cos^2 \theta' + \sin^2 \theta' \cos^2 \phi'),$$

and again using the fact that the density does not depend on ϕ' :

$$A = \int dr' \int_0^\pi d\theta' \sigma \int_0^{2\pi} d\phi'\, r'^2 \sin \theta'\, r'^2(\cos^2 \theta' + \sin^2 \theta' \cos^2 \phi').$$

Since

$$\int_0^{2\pi} \cos^2 \phi'\, d\phi' = \pi,$$

$$A = 2\pi \int dr' \int_{-1}^{+1} d(\cos \theta') \sigma r'^4(\cos^2 \theta' + \tfrac{1}{2} \sin^2 \theta'),$$

and so

$$C - A = -2\pi \int dr' \int_{-1}^{+1} d(\cos \theta') \sigma r'^4 \tfrac{1}{2}(3 \cos^2 \theta' - 1).$$

Now let us calculate the attraction of the body at some point P outside it, having co-ordinates r, θ, ϕ (fig. 36). The potential at this point will be G times the integral of each element of mass in the body divided by the distance D between it and the exterior point. Let χ be the angle between the lines joining the point P and an element of mass in the body to the centre of mass of the body. Then by the cosine rule :

$$D^2 = r^2 + r'^2 - 2rr' \cos \chi.$$

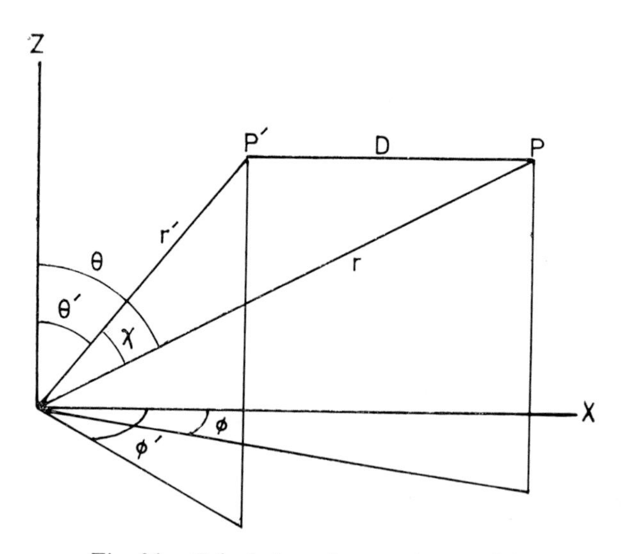

Fig. 36. Calculation of external potential.

Hence the potential is equal to :

$$-G \int_V \frac{\sigma \, dv}{(r^2 + r'^2 - 2rr' \cos \chi)^{\frac{1}{2}}}$$

$$= -\frac{G}{r} \int_V \sigma \, dv \left(1 + \frac{r'^2}{r^2} - \frac{2r'}{r} \cos \chi\right)^{-\frac{1}{2}}$$

$$= -\frac{G}{r} \int_V \sigma \, dv \left\{1 + \frac{r'}{r} \cos \chi + \frac{r'^2}{2r^2}\left(3 \cos^2 \chi - 1\right) + \ldots\right\}.$$

Now $\int_V \sigma \, dv$ is just the mass, M, of the body and so the first term in the integral is $-GM/R$, the same as for a spherically symmetrical body.

The next term is :

$$-\frac{G}{r^2} \int_V dv \, \sigma r' \cos \chi.$$

Now by a result in spherical trigonometry,

$$\cos\chi = \cos\theta\cos\theta' + \sin\theta\sin\theta'\cos(\phi-\phi'),$$

and so the term becomes :

$$\frac{G}{r^2}\int_V dv\,\sigma r'\{\cos\theta\cos\theta' + \sin\theta\sin\theta'\cos(\phi-\phi')\}.$$

But the origin of co-ordinates is taken at the centre of mass of the body and therefore

$$\int_V dv\,\sigma r'\cos\theta'$$

is zero while

$$\int_V dv\,\sigma r'\sin\theta'\cos(\phi-\phi')$$

vanishes by the same argument or because the body is symmetrical about the polar axis.

The last term to consider is :

$$-\frac{G}{2r^3}\int_V dv\,\sigma r'^2(3\cos^2\chi - 1).$$

Now

$$\cos^2\chi = \cos^2\theta\cos^2\theta' + 2\cos\theta\cos\theta'\sin\theta\sin\theta'\cos(\phi-\phi') + \sin^2\theta\sin^2\theta'\cos^2(\phi-\phi').$$

Again, let us use the fact that σ is independent of ϕ' to integrate with respect to ϕ'. Remembering that

$$\int_0^{2\pi}\cos\phi'\,d\phi' \quad\text{and}\quad \int_0^{2\pi}\sin\phi'\,d\phi'$$

are zero and that

$$\int_0^{2\pi}\cos^2\phi'\,d\phi' \quad\text{and}\quad \int_0^{2\pi}\sin^2\phi'\,d\phi'$$

are π, it follows that

$$\int_0^{2\pi}\cos^2\chi\,d\phi' = 2\pi(\cos^2\theta\cos^2\theta' + \tfrac{1}{2}\sin^2\theta\sin^2\theta')$$

and hence that

$$-\frac{G}{2r^3}\int_V dv\,\sigma r'^2(3\cos^2\chi - 1)$$

$$= -2\pi\frac{G}{4r^3}\int dr'\int_{-1}^{+1} d(\cos\theta')\sigma r'^4(3\cos^2\theta' - 1)(3\cos^2\theta - 1).$$

The potential is therefore :

$$-\frac{GM}{r} + \frac{G(C-A)}{2r^3}(3\cos^2\theta - 1) + \text{terms of higher powers of } r,$$

that is :

$$-\frac{\mu}{R_e}\cdot\left(\frac{R_e}{r}\right)\left[1 - \frac{C-A}{MR_e^2}\left(\frac{R_e}{r}\right)^2\tfrac{1}{2}(3\cos^2\theta - 1) + \ldots\right],$$

where R_e is a constant, usually the equatorial radius of the Earth.

This important result is known as *McCullagh's theorem.* Its practical importance lies in the fact that while $(C-A)/MR_e^2$ is about one in one thousand for the Earth, all other variations of the potential, which are proportional to higher powers of r, are still smaller by a factor of more than a thousand. Furthermore, as will be seen in Chapter 5, the difference between the polar and equatorial radii of the Earth is directly related to $(C-A)$.

Satellites around a flattened Earth

We can now go on to work out the effect on the orbit of a satellite of a part of the potential proportional to $(C-A)/MR_e^2$. We have to

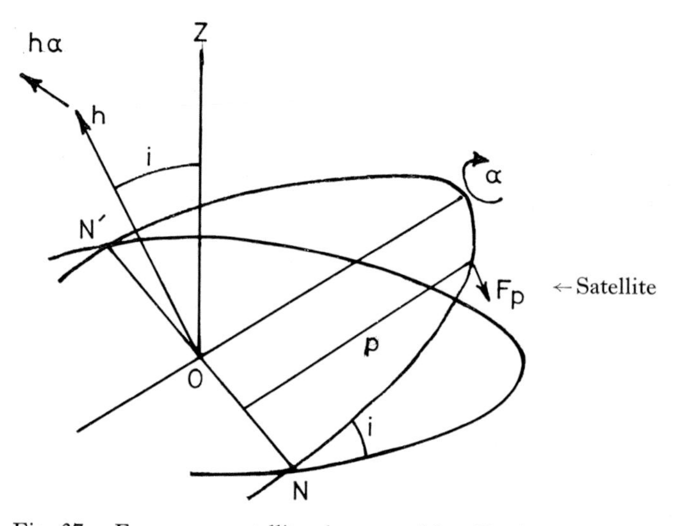

Fig. 37. Force on a satellite about an oblate Earth.

calculate the potential and the force on a satellite that moves in an elliptical orbit, that is to say, the co-ordinates r and θ must be those of a point on such an orbit. To simplify things, let us suppose that the orbit is in fact circular. Since the focus is at the centre of mass of the attracting body (the Earth), r stays constant and so when we differentiate the potential to find the force we do so only with respect to the angle θ. We shall see that the force that results has a component perpendicular to the plane of the orbit, a force that exerts a couple on the satellite about the line of nodes. A couple is equal to a rate of change of angular momentum but since the total angular momentum of the satellite remains constant, the component in the direction of the line of nodes can alter only if the direction of total angular momentum, which is perpendicular to the plane of the orbit, also changes. The required change is a rotation of the plane about the line in the plane perpendicular to the line of the nodes. The situation is shown by the diagram in fig. 37. NN' is the line of the nodes, p is the perpendicular distance of the satellite from the line of the nodes and F is the force on the satellite due to the part of the potential proportional to $C-A$. F_p is the component of that force perpendicular to the plane of the orbit and the couple about the line of the nodes is therefore pF_p. Let α be the rate at which the orbit rotates about the line perpendicular to the line of the nodes. If h is the total angular momentum of the satellite about the centre of attraction, the rate at which angular momentum is generated in the direction of the line of the nodes is $h\alpha$ and hence

$$pF_p = -h\alpha.$$

Now if the satellite orbit is inclined to the equator at the angle i, the rotation at the rate α corresponds to a rotation of the line of the nodes along the equator at the rate $\alpha/\sin i$ and hence $\dot{\Omega}$, the rate of rotation of the nodes, is given by

$$\dot{\Omega} = -\frac{pF_p}{h \sin i}.$$

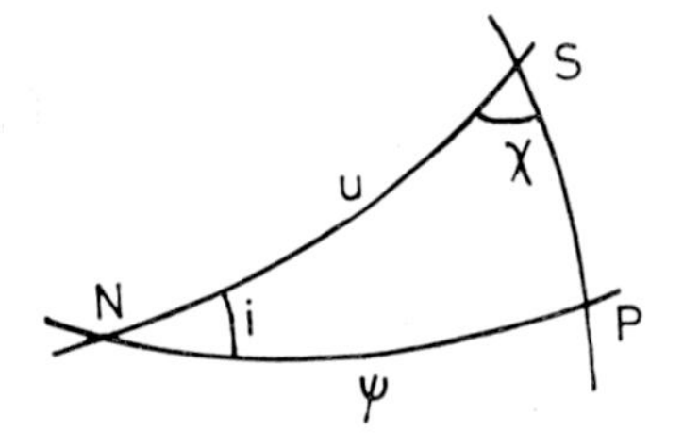

Fig. 38. Geometry of a satellite orbit.

To calculate pF_p we must express θ in terms of the inclination of the orbit and the position of the satellite in it. Let the position of the satellite, S, be given by the angle u measured in the plane of the orbit from the ascending node. (Since we have supposed the orbit to be circular, we do not have to know where perigee is.) Let the plane of the meridian that passes through the satellite cut the equator at P and let the angle between P and the ascending node N be ψ (fig. 38). Let the angle between the meridian and the orbit at S be χ. SP is equal to $\frac{1}{2}\pi - \theta$. From spherical trigonometry, we find that

$$\cos\theta = \sin i \sin u.$$

Further, from the sine rule in the appropriate spherical triangles :

$$\frac{\sin \chi}{\sin \psi} = \frac{\sin i}{\cos \theta} = \frac{1}{\sin u},$$

and

$$\frac{\sin \psi}{\sin u} = \frac{\cos i}{\sin \theta}.$$

Hence

$$\sin \chi = \frac{\cos i}{\sin \theta}.$$

Let us rewrite the $(C-A)$ term in the potential as :

$$V_2 = \frac{\mu}{r}\left(\frac{R_e}{r}\right)^2 J_2 \tfrac{1}{2}(3 \cos^2 \theta - 1),$$

where we have put :

$$J_2 = \frac{C-A}{MR_e^2}.$$

The force tangential to the meridian is found by differentiating the potential V_2 with respect to θ and is $-\partial V_2/r\partial\theta$, or

$$\frac{3\mu}{r^2}\cdot\left(\frac{R_e}{r}\right)^2 J_2 \cos \theta \sin \theta.$$

The component F_p perpendicular to the plane of the orbit is :

$$\frac{3\mu}{r^2}\left(\frac{R_e}{r}\right)^2 J_2 \cos \theta \sin \theta \sin \chi.$$

Substituting for $\cos \theta$ and $\sin \chi$ we find that F_p is :

$$\frac{3\mu}{r^2}\left(\frac{R_e}{r}\right)^2 J_2 \sin i \cos i \sin u.$$

The perpendicular p is :

$$r \sin u,$$

and therefore the couple about the line of nodes, pF_p, is :

$$\frac{3\mu}{r}\left(\frac{R_e}{r}\right)^2 J_2 \sin i \cos i \sin^2 u.$$

Using the expression $-pF_p/h \sin i$ for the rate of rotation of the lines of the nodes, it follows that

$$\dot{\Omega} = -\frac{3\mu}{hr}\left(\frac{R_e}{r}\right)^2 J_2 \cos i \sin^2 u,$$

or, since $h^2 = r\mu$ and $\mu^{\frac{1}{2}} r^{-3/2} = 2\pi/T$,

$$\dot{\Omega} = -\frac{2\pi}{T}\left(\frac{R_e}{r}\right)^2 3J_2 \cos i \sin^2 u.$$

This is the instantaneous rate of rotation. The change in the course of one orbit is found by integrating with respect to u from 0 to 2π and since

$$\int_0^{2\pi} \sin^2 u \, du = \pi,$$

the change is :

$$\delta\Omega = -2\pi n\left(\frac{R_e}{r}\right)^2 . \tfrac{3}{2}J_2 \cos i,$$

or

$$\dot{\Omega}_{\text{av}} = -\tfrac{3}{2}n\left(\frac{R_e}{r}\right)^2 J_2 \cos i,$$

where

$$n = 2\pi/T.$$

The actual behaviour of the orbits of artificial satellites is very close to that given by this formula but if the rates of rotation of the nodes of artificial satellites are compared for satellites with different inclinations, it is found that there are slight differences from the formula. More complete calculations show that additional terms in the potential that we have neglected also give rise to steady changes in the position of the nodes. Before explaining what these are, a second effect of the potential term proportional to J_2 must be mentioned. The force F has a component in the plane of the orbit which gives a couple that averages out to zero in a circular orbit but that has an average different from zero in an elliptical orbit. The effect of this couple is to cause the position of perigee to move round the orbit at an average rate equal to

$$-\tfrac{3}{4}nJ_2\left(\frac{R_e}{r}\right)^2 (5\cos^2 i - 1).$$

The observed behaviour of artificial satellites

The general term in the potential that does not depend on ϕ has the form :

$$\frac{GM}{r}\left(\frac{R_e}{r}\right)^n J_n P_n(\cos\theta),$$

where $P_n(\cos\theta)$ is a polynomial in $\cos\theta$, the highest power of which is $\cos^n\theta$. When n is even all the terms of $P_n(\cos\theta)$ are even powers of $\cos\theta$ and when n is odd they are all odd. The number of times that $P_n(\cos\theta)$ passes through zero as θ varies from 0 to π is n. $P_2(\cos\theta)$ is $\tfrac{1}{2}(3\cos^2\theta - 1)$, an expression we have encountered above.

These polynomials are known as *Legendre polynomials* and are special examples of certain functions of position on the surface of a sphere called *surface harmonics*. The Legendre polynomials are *zonal harmonics* and depend only on latitude. *Sectorial harmonics* depend only on longitude and *tesseral harmonics* depend on both. The mathematical properties they possess make them especially useful in expressing the way in which quantities vary over a sphere, just as sine and cosine waves express the variation of a quantity around a circle. Some examples are shown in fig. 39.

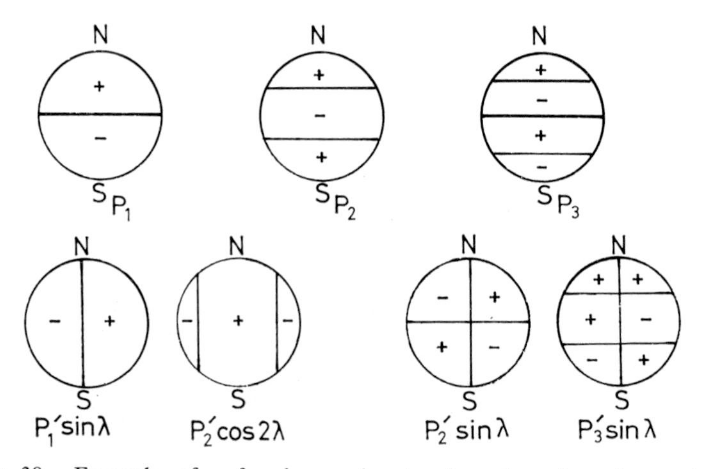

Fig. 39. Examples of surface harmonics showing where they change sign.

All the terms in the potential with even n give rise to steady rates of change of the line of nodes, while those with odd n cause the position of the line of nodes to oscillate about its mean position with a period equal to the rate at which perigee rotates in the orbit as a result of the J_2 term in the potential. These and certain other effects can be detected by careful observations of the positions of satellites and certain of the coefficients J_n have been estimated from the observations. The effects of the J_2 term can be seen in the motion of our natural satellite, the Moon, but because the effects are proportional to $(R_e/r)^n$, close satellites, for which (R_e/r) is only slightly less than 1, must be observed if the higher-order terms are to be detected.

Some values of the coefficients J_n are given in Table 1.

Gravity also varies with longitude ϕ but it is much more difficult to observe the corresponding changes in the orbits and although some variations with ϕ are reasonably well established, they are in general much less well known than those variations that do not depend on longitude.

Table 1. Values of the coefficients J_n in the potential of the Earth

n	$10^6 J_n$
2	1082·7
3	−2·6
4	−1·6
5	−0·1
6	+0·7
7	−0·5
8	0
9	0·1
10	−0·1
11	0·3
12	−0·3

Problems

4.1. The period of an artificial satellite of the Earth is 92·5 min and the semi-major axis of its orbit is 6800 km. What is the mass of the Earth? (6×10^{24} kg.)

$$G = 6{\cdot}67 \times 10^{-11}\ \mathrm{Nm^2/kg^2}.$$

If the mass of the Moon is 1/81 times that of the Earth, what is the period of a satellite about the Moon in an orbit of semi-major axis 2000 km? (129 min.)

4.2. From the following data for Sputnik 2, calculate the rate of regression of its node:

period	98 min,
semi-major axis	7000 km,
inclination	65·3°.

Take J_2 to be $1{\cdot}083 \times 10^{-3}$ and the equatorial radius of the Earth to be 6400 km. (3°/day.)

4.3. Calculate J_2 for Mars from the following data:

Satellite Phobos:	period	0·32 days,
	semi-major axis	9000 km,
	inclination	1·1°,
	rate of regression of node	0·56°/day,
	equatorial radius of Mars	3375 km.

($2{\cdot}0 \times 10^{-3}$.)

CHAPTER 5

the size, shape and structure of the Earth

The mass and size of the Earth

WE have seen in the previous chapters that there are two ways in which we can find the value of gravity in the neighbourhood of the Earth—we may measure it directly on the surface, or we may observe the way in which it controls the movement of a satellite of the Earth. These two ways must give the same answer and by comparing them, it is possible to find the size of the Earth. Let us make the simplest possible assumption, that the Earth is a sphere of radius R, and let the value of gravity on the surface be g. Let the distance of the Moon be $R_☾$ and let the time she takes to make one revolution in her orbit be T. Again let us make the simple assumption that the orbit of the Moon is a circle, so that the acceleration of the Moon towards the Earth is:

$$R_☾(2\pi/T)^2,$$

since the angular velocity of the Moon in her orbit is $(2\pi/T)$.

But this acceleration must be equal to the gravitational attraction of the Earth at the distance of the Moon and that, by the inverse square law, will be equal to:

$$g(R/R_☾)^2.$$

We therefore find that

$$R^2 = \frac{R_☾^3}{g}\left(\frac{2\pi}{T}\right)^2,$$

an equation from which R may be calculated if we know T, g and $R_☾$.

Kepler's formula may also be used to find the masses of double stars. The period of rotation of two stars around each other can often be found from the variation in the total light received from them on Earth. We often cannot observe the linear separation of the stars, but it is possible to find the change of velocity in the orbit. If the period is T, the angular velocity is $2\pi/T$ and the change of linear velocity Δv in the direction of the Earth on going round the orbit is

$$2(2\pi/T)\,a.$$

Kepler's formula gives us:

$$\mu = \frac{4\pi^2\,a^3}{T^2} = \frac{T}{16\pi}(\Delta v)^3.$$

Each star moves round the common centre of mass and if the two masses, m_1 and m_2 are comparable, μ is equal to $G(m_1+m_2)$.

But why should we make observations of the Moon to find out how big the Earth is? Let us arrange the various quantities in the order of the accuracy with which they can be measured. The most accurate is T, the period of the Moon in her orbit, for that can be found from observations of the Moon over very many periods. The next most accurate is the distance of the Moon from the Earth which is now known to about one part in a million as a result of radar measurements. We have seen that the value of gravity at any one place can be measured to much better than one part in a million, but the value of gravity that we need to use in the calculations we are now thinking of is the average value of gravity over the whole surface of the Earth, and that, because of the variations of gravity, is not known to better than two or three parts in a million. So we might expect to calculate the size of the Earth from surface gravity and observations of the Moon to about the same accuracy of two or three parts in a million.

Defining the shape of the Earth

By comparison, how accurately is the size of the Earth known from measurements on the surface? What in fact, is meant by the size of the Earth, by its mean radius? We know that the surface of the solid Earth is very irregular and that it is described by drawing maps that show the height of the surface above sea level, but how can we know the shape of the surface of the sea which is continually in movement with the waves and tides, and on which no marks can be fixed to allow measurements of its shape. The study of gravity enables the shape of the sea surface to be calculated. Imagine the sea undisturbed by the tides and by waves. Suppose that the surface is such that at some parts the potential of gravity is greater than at others. This means that since water can flow freely, it will do so until at no place is there water that can move to a place of lower gravitational potential. The surface of the sea at rest would therefore be one over which the potential is a constant. There is the slight complication that the sea rotates with the Earth and so the surface is at rest under the combined action of the rotation and of the gravitational attraction of the matter of the Earth. This, then is what is meant by the shape of the sea-level surface—it is the shape of that surface over which the potential is constant and equal to the average value at the sea surface, and when we speak of the shape and size of the Earth, it is of this surface that we think. The radius of the equator of this surface is what is meant by the equatorial radius of the Earth and the polar flattening of this surface is the polar flattening of the Earth. These quantities can be calculated from the known variation of gravity outside the Earth.

The shape, gravity and potential of the rotating Earth

In the first place, the effect of the rotation of the Earth must be considered. In fig. 40, let P be a point on the surface of the Earth with polar co-ordinates (r, θ), where r is measured from the centre of the Earth and θ is measured from the direction of the North Pole. The perpendicular distance of P from the polar axis, the axis of rotation, is therefore $r \sin \theta$ and the acceleration of P equivalent to the rotation of the Earth with angular velocity ω is $\omega^2 r \sin \theta$ in the direction perpendicular to the polar axis and directed outwards away

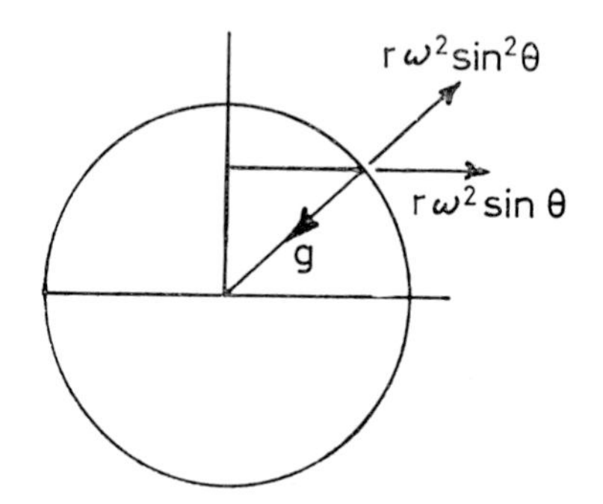

Fig. 40. Gravity on the rotating Earth.

from it. Now the attraction of gravity at P is along the direction joining P to the centre of the Earth and the outward component of the rotational acceleration in that direction is $r\omega^2 \sin^2 \theta$. This is the same as $-(\partial/\partial r)(-\frac{1}{2}r^2 \omega^2 \sin^2 \theta)$ and so the rotational acceleration can be considered to be derived from a potential equal to $-\frac{1}{2}r^2 \omega^2 \sin^2 \theta$.

If we call V the potential of the gravitational attraction of the matter of the Earth, the total potential at the surface of the sea will be :

$$V - \tfrac{1}{2}r^2 \omega^2 \sin^2 \theta.$$

If very small terms are ignored, V is equal to :

$$-\frac{\mu}{r}\left[1 - J_2\left(\frac{a}{r}\right)^2 \tfrac{1}{2}(3 \cos^2 \theta - 1)\right]$$

and the shape of the sea level surface will be such that

$$-\frac{\mu}{r}\left[1 - J_2\left(\frac{a}{r}\right)^2 \tfrac{1}{2}(3 \cos^2 \theta - 1)\right] - \tfrac{1}{2}r^2\omega^2 \sin^2 \theta$$

is constant on it.

We know that the shape of the Earth is very nearly spherical, so let us write r, the distance of P from the centre of the Earth, as $a\{1 + \epsilon(\theta)\}$ where $\epsilon(\theta)$ is a small quantity that varies with the angle θ. Now substitute this expression for r in the formula for the potential and equate the result to a constant :

$$-\frac{\mu}{a}\{1-\epsilon(\theta)\}[1-\tfrac{1}{2}J_2(3\cos^2\theta-1)]-\tfrac{1}{2}a^2\omega^2\sin^2\theta = C,$$

where such small quantities as ϵ^2, ϵJ_2, $\epsilon\omega^2 a^2$, are ignored.

The result will be a constant only if the potential on the surface is not dependent on θ. The parts of the potential that do depend on θ are :

$$\frac{\mu}{a}\{\epsilon(\theta)+\tfrac{3}{2}J_2\cos^2\theta\}+\tfrac{1}{2}a^2\omega^2\cos^2\theta$$

and so

$$\epsilon(\theta) = -\tfrac{1}{2}(3J_2+\mathbf{m})\cos^2\theta,$$

where

$$\mathbf{m} = a^3\omega^2/\mu.$$

This means that the shape of the sea-level surface is given by :

$$r = a(1-f\cos^2\theta),$$

f being $\tfrac{1}{2}(3J_2+\mathbf{m})$.

Equation of an ellipse referred to its centre

This expression is the equation for an ellipse in polar co-ordinates referred to the centre. For let the equation in Cartesian co-ordinates with the origin at the centre of the ellipse be :

$$\frac{x^2}{a^2}+\frac{y^2}{b^2} = 1.$$

With θ measured from the minor axis :

$$x = r\sin\theta, \quad y = r\cos\theta,$$

and so if $b = a(1-f)$

$$r^2(1-\cos^2\theta)+\frac{r^2\cos^2\theta}{(1-f)^2} = a^2,$$

or

$$r^2(1-\cos^2\theta)+r^2\cos^2\theta(1+2f) = a^2,$$

that is :

$$r^2(1+2f\cos^2\theta) = a^2,$$

so that

$$r = a(1-f\cos^2\theta),$$

neglecting quantities like f^2 and smaller.

From this it will be seen that f is the relative difference between the semi-major and semi-minor axes :

$$f = \frac{a-b}{a},$$

a quantity called the *polar flattening* of the Earth.

The variation of gravity

The quantity $\mathbf{m}$ is equal to the ratio of the rotational acceleration at the equator ($a\omega^2$) to the gravitational acceleration there, μ/a^2, and is about $3{\cdot}45 \times 10^{-3}$. Since J_2 is about $1{\cdot}0827 \times 10^{-3}$, f is $3{\cdot}35 \times 10^{-3}$ or $1/298{\cdot}5$. Since the radius of the Earth is 6400 km, the difference, $(a-b)$ is just over 20 km.

That the Earth is flattened at the poles has of course been known since long before the advent of artificial satellites and indeed Newton gave an estimate of f which is not so very far from that now accepted. The flattening of the Earth could be found in two ways, one by direct measurement over the solid surface and the other by measurement of gravity over the surface ; both these methods are inferior to the determination from the potential derived from satellite observations because it is difficult to obtain representative average values from what are still rather sparse observations at the surface of the Earth.

Taking the potential at a distance r from the centre of the Earth to be :

$$-\frac{\mu}{r}\left[1-J_2\left(\frac{a}{r}\right)^2 \tfrac{1}{2}(3\cos^2\theta-1)\right]-\tfrac{1}{2}r^2\,\omega^2(1-\cos^2\theta),$$

the value of gravity at a point rotating with the Earth is :

$$-\frac{\partial}{\partial r}\left[-\frac{\mu}{r}\left\{1-J_2\left(\frac{a}{r}\right)^2 \tfrac{1}{2}(3\cos^2\theta-1)\right\}-\tfrac{1}{2}r^2\,\omega^2(1-\cos^2\theta)\right].$$

At the sea-level surface, r is $a\,(1-f\cos^2\theta)$ and so by substituting this value of r in the expression for gravity, the value of gravity at sea level is found to be :

$$\frac{\mu}{a^2}\left[1+2f\cos^2\theta-\tfrac{3}{2}J_2(3\cos^2\theta-1)-\mathbf{m}(1-\cos^2\theta)\right]$$
$$= g_0[1+(2f-\tfrac{9}{2}J_2+\mathbf{m})\cos^2\theta],$$

where

$$g_0 = \frac{\mu}{a^2}(1+\tfrac{3}{2}J_2-\mathbf{m}).$$

This is of the form we have met previously namely :

$$g = g_0(1+\beta\cos^2\theta),$$

with

$$\beta = 2\mathbf{m} - \tfrac{3}{2}J_2,$$

or

$$\beta = \tfrac{5}{2}\mathbf{m} - f.$$

β was found from measurements of gravity over the Earth, and since $\mathbf{m}$ is easily calculated from the speed of rotation of the Earth, f could be derived. Nowadays, J_2 is found from satellite observations and β is derived and used in calculating gravity anomalies. The value of f found from surface gravity observations was about 1/296 but was some twenty times less accurate than the value now derived from satellite data.

The flattening from survey measurements

The flattening can also be found from measurements of the length of an arc of the meridian at the surface and this was the earliest way in which it was determined. The length of an arc that subtends an angle $d\theta$ is $rd\theta$ neglecting quantities of order f^2; survey measurements over the surface give the length of the arc, while from observations of the stars made at the two ends of the arc, the angle $d\theta$ can be obtained. A combination of the two measurements gives r and by repeating the measurements at sufficiently different latitudes, f can be found. Values of f derived in this way show quite large variations, mainly because local variations of gravity cause the observed directions of the stars to differ quite considerably from those to be expected on a surface with a simple ellipsoidal shape.

The Metric system, now generally used for almost all measurements in most countries of the world, was developed as the result of early measurements of the flattening of the Earth. Newton had predicted that the Earth should be flattened at the poles and had estimated the degree of flattening, as already mentioned, but a French scientist, Cassini, considered that the poles were elongated with respect to the equator rather than flattened. To settle the issue, the French Academy organized two expeditions, one to Lapland and one to Peru, to measure the length of an arc subtending a degree of latitude, the one near the north pole and the other close to the equator. Out of this work, which confirmed Newton's ideas, grew the concept of a standard of length based on the size of the Earth.

The geoid

We know that gravity at the surface of the Earth does not obey the simple formula for an ellipsoid of rotation and we know that satellite observations show that the external potential contains many more terms than the one proportional to J_2. The surfaces of constant potential cannot therefore be simple ellipsoids of revolution but must be more complex. Suppose that the potential at some point exceeds the simple ellipsoidal value by an amount δV and suppose that the equipotential surface lies at a height N above the ellipsoidal surface. The gradient of potential over the height N is $\delta V/N$ and this must be equal to the observed value of gravity at the point. Thus

$$N = -\frac{\delta V}{g},$$

a formula that enables the shape of the equipotential surface relative to an ellipsoid to be found from the corresponding departures of the potential, δV. The surface on which the actual potential is constant is called the *geoid* and a map of the value of N, the elevation of the geoid above an ellipsoid with a specified value of f or J_2, is a convenient way of representing the potential. Such a map is shown in fig. 41.

The internal structure of the Earth

The shape of the sea-level surface and the value of gravity upon it are uniquely determined by the external gravitational potential, and the size of the Earth, that is to say, the equatorial radius of the sea-level surface, can be found from one measured dimension, such as the distance of the Moon. But while the shape and external gravity are fixed by the external potential, the potential can be produced by any of a wide range of distributions of matter within the Earth. Thus only rather general knowledge of the interior of the Earth can be obtained from a knowledge of gravity, and indeed it is only because the Earth is not quite spherically symmetrical that gravity can tell us anything. Were the density in the Earth a function of radius alone, so that the Earth could be considered to be made up of a succession of concentric spherical shells, the attraction of each would be just the same as that of a point mass at the common centre, gravity would be uniform over the surface, and nothing could be said about the way in which the density varied with radius. Furthermore, the small variations of gravity over the surface can provide only very general information about the internal distribution of mass unless they are combined with other data; in that case, the inferences may be very important.

The moments of inertia of the Earth and the central core

The moments of inertia of the Earth affect gravity only through the difference $(C-A)$, to which the coefficient J_2 in the external potential is proportional. The rate of precession of the Earth enables C and A to be found separately. The axis of rotation of the Earth precesses

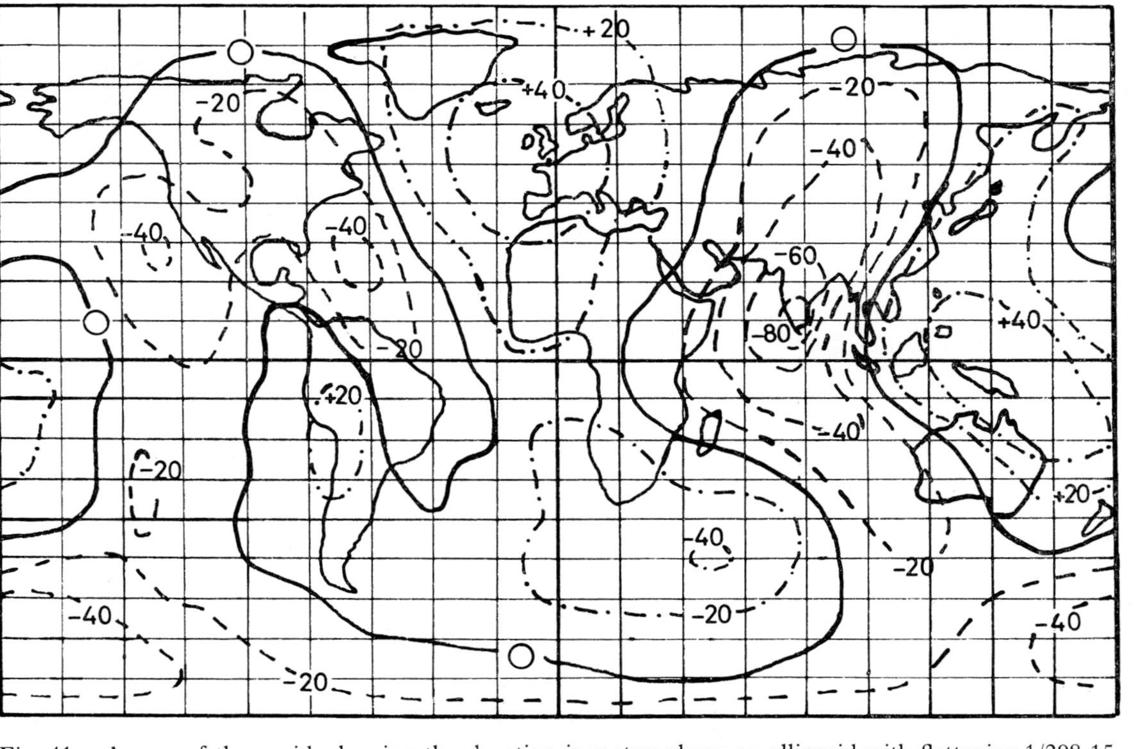

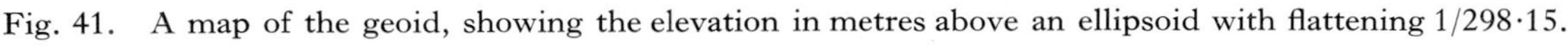
Fig. 41. A map of the geoid, showing the elevation in metres above an ellipsoid with flattening 1/298·15.

because the attraction of the Moon acting on the equatorial bulge of the Earth produces a couple about an axis in the equatorial plane tangential to the orbit of the Moon (the attraction of the Sun produces a similar couple). The couple is proportional to $(C-A)$, while the angular momentum of the Earth about its polar axis is $C\omega$, where ω is the spin angular velocity of the Earth. The angular velocity of the Earth about an axis in the plane of the orbit and normal to the orbit is therefore proportional to $(C-A)/C$ and it is this rotation, which, averaged over the orbit of the Moon, gives rise to the steady precession of the Earth (the periodic component is the nutation). The observed precession may therefore be used to find the quantity H which is equal to $(C-A)/C$. Since J_2 is equal to $(C-A)/Ma^2$, C/Ma^2 is equal to J_2/H. The numerical value of H is $3{\cdot}275 \times 10^{-3}$ and so C is equal to $0{\cdot}3306Ma^2$.

The moment of inertia of a sphere of constant density and radius a is $0{\cdot}4Ma^2$, while that of a thin spherical shell is $\frac{2}{3}Ma^2$ and that of a mass point at the centre is zero. These figures suggest that the density of the Earth increases towards the centre and more detailed calculations show that that is indeed so. As an example, consider the Earth to be comprised of two zones, an inner high density zone of radius equal to half the radius of the Earth, and an outer low density zone between the inner zone and the surface. Table 2 shows the values of C/Ma^2 for different values of the ratio of the densities in the two zones. The value of C/Ma^2 is not very sensitive to the ratio of densities and it is clear that the density towards the centre of the Earth must be much greater than that near the surface. Further than this it is not possible to go with gravitational information alone, but by combining the mass and moments of inertia of the Earth with knowledge of the speed of elastic waves in the Earth at different depths as found from the times of travel of waves from earthquakes, the variation of density with radius within the Earth can be worked out. It is then found that the Earth is divided into a number of distinct zones. The innermost is the *inner core*, which is believed to be solid and which has a radius of 1250 km. Next comes the liquid *core* with a radius of 3470 km and a density of about 10 000 kg/m³. Between

Table 2. Values of I/Ma^2 for different degrees of condensation

Ratio of inner to outer density	I/Ma^2
2	0·367
3	0·339
Earth	0·330
4	0·318

the core and a radius of 5370 km is the *lower mantle* with a density of 5000 kg/m³ and outside that is the *upper mantle* with a density of around 4000 kg/m³ and extending to within some 30 km of the surface. The outermost layers are the crust, the continents and the oceans which have already been described in Chapter 3.

Evidently a major division in the Earth is that between the mantle and the core and it is interesting to see if any variations of gravity can be related to the boundary between these two regions, particularly since the change of density across that boundary is greater than anywhere else in the Earth. Unfortunately it is not possible to answer this question at present, although we can show that any variations of gravity at the surface which change sign within a distance of about 30° must be due to variations of density above the core. There may be such rapid variations of density within the core but they would not appear in the values of gravity at the surface. Since we know from satellite observations and from surface measurements of gravity that there are such rapid variations at the surface, we conclude that there are corresponding rapid variations of density in the mantle, but without additional information nothing more can be said about them. Some of the observed variations of gravity may be due to irregularities of the boundary between the core and the mantle, and careful studies of the times of travel of waves from earthquakes may enable the shape of that boundary to be worked out. The problem is important because irregularities in the boundary between the fluid core and the solid mantle would probably give rise to fluid currents in the core which would affect the magnetic field of the Earth : the field is almost certainly the consequence of currents in the electrically conducting core.

The strength of the Earth

If the Earth were wholly fluid and were at rest apart from spinning on its axis, the density could at most vary with the distance from the centre and the only variation of gravity over the surface would be that corresponding to the polar flattening, flattening which would be the consequence of the spin. Such an Earth would be said to be in hydrostatic equilibrium for the liquid of which it was composed would be rotating as a whole without any relative internal motions. Since gravity shows small variations other than that corresponding to the flattening, it is clear that the Earth is not in hydrostatic equilibrium†, and either it is effectively fluid throughout but maintains relative motions, or else it is in part solid and the solid parts are strong enough to support the differences of density that give rise to the variations of gravity. Since

†The most obvious evidence that the Earth is not in hydrostatic equilibrium is that we live on solid ground and that the seas do not cover everything uniformly.

we see that at least the outer parts of the Earth are solid, the second seems at first sight the more likely situation. We do not know where in the Earth the differences of density are to be found that give rise to the observed variations of gravity, so that it is not possible to say definitely how strong the material of the Earth must be if it is to support those differences against the forces exerted on them by the gravitational attraction of the rest of the Earth. It is, however, possible to say what the least strength must be by choosing the distribution of density in such a way that the elastic energy stored in resisting the gravitational attraction is as small as possible. On that basis, the strength of the solid mantle of the Earth must be about $1{\cdot}5 \times 10^7$ N/m^2 (150 atm).

However, although the outer parts of the Earth appear to us to be solid, they may be like pitch, resistant to short sharp shocks but able to flow under forces applied for a very long time. We know that heat is flowing steadily out of the surface of the Earth and so it has been suggested that there may be convection currents in the mantle, with material which is somewhat hotter and therefore less dense than the average rising towards the surface, while in other places, material which is cooler and therefore more dense is moving downwards. It is very difficult at present to understand such possible currents because although quite a lot is known about convection in ordinary liquids like water, almost nothing is known about the behaviour of a material which is solid if considered for a short time but fluid over a very long time. There is, however, a great deal of evidence that the continents have moved about in the course of the life of the Earth with the formation of the Atlantic and other oceans, and an obvious possibility is that the continents have been carried on convection currents in the upper mantle. If these suppositions are correct, the variations of gravity outside the Earth and the corresponding irregularities would be expected to be some guide to the way in which the currents were flowing.

Problems

5.1. The Moon completes her orbit about the Earth in 27·5 days. Her distance from the Earth is 384 400 km. If the mean value of gravity on the surface of the Earth is 9·79 m/s^2, what is the mean radius of the Earth? ($6{\cdot}37 \times 10^3$ km.)

5.2. Given that the equatorial radius of the Earth is 6378 km and that the value of gravity at sea level at the equator is 9·7804 ms^{-2}, calculate the parameter **m**. ($3{\cdot}45 \times 10^{-3}$.)

If $J_2 = 1{\cdot}0827 \times 10^{-3}$ calculate f, $1/f$ and β. ($3{\cdot}35 \times 10^{-3}$, 1/298·5, $5{\cdot}28 \times 10^{-3}$.)

5.3. The value of the coefficient J_3 in the potential is $2{\cdot}5 \times 10^{-6}$. What is the amplitude of the corresponding undulation of the geoid? (16 m.) (Use the formula, $N = -\delta V/g$ taking δV to be the third harmonic with coefficient J_3.)

CHAPTER 6

inconstant gravity

In the previous chapters it has been quietly assumed that gravity at any point on the surface of the Earth is always the same. It is certainly the case that any changes there may be in the course of time are very small, but it is also certain that some occur. The total attraction of gravity comprises not only the attraction of the matter of the Earth itself, but also the attractions of the Sun and the Moon, which are continually changing with the position of the Sun and the Moon in the sky. The effects of these changes are somewhat indirect because the Earth yields elastically under the changing value of gravity, and the observed effects, when compared with the calculated attractions of the Sun and Moon, add to our knowledge of the interior of the Earth. The more speculative question, and in some ways the more interesting one, is whether the attraction of the Earth itself is changing, but that is a much more difficult question to answer.

The attraction of the Sun and the Moon

In fig. 42, let P be the position of a point on the surface of the Earth at a distance a from the centre, and let M be the Moon. If R

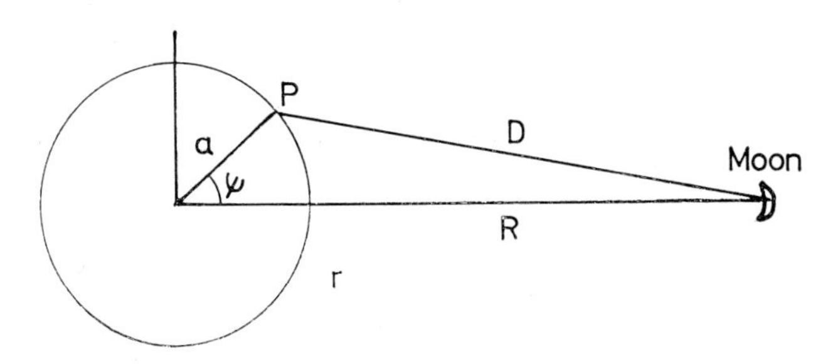

Fig. 42. Potential of the Moon at the Earth.

is the distance of the Moon from the centre of the Earth and if D is the distance of the Moon from P, then

$$D^2 = a^2 + R^2 - 2aR \cos \psi,$$

where ψ is the angle seen from the centre of the Earth between the direction of P and M. If the mass of the Moon is also denoted by M, the potential of the attraction of the Moon at P is:

$$-\frac{GM}{D}.$$

If the reciprocal distance, $1/D$, is expanded by the binomial theorem as was done in Chapter 4, the potential may be written as:

$$-\frac{\mu}{R}\left[1+\frac{a}{R}\cos\psi+\left(\frac{a}{R}\right)^2\tfrac{1}{2}(3\cos^2\psi-1)+\ldots\right].$$

The first term is a constant and so produces no force at P. The second term corresponds to a force:

$$-\frac{\partial}{\partial a}\left[-\frac{\mu a}{R^2}\cos\psi\right] = \frac{\mu}{R^2}\cos\psi$$

along the line joining the centre of the Earth to P. Now the acceleration of the centre of mass of the Earth towards the Moon due to the attraction of the Moon is:

$$\mu/R^2$$

and the radial component of that force at P is $(\mu/R^2)\cos\psi$, just the force derived from the second term in the potential of the Moon at P. But this centre of mass acceleration is that of the motion of the Earth in its orbit about the Moon, or more strictly, in its orbit about the common centre of mass of the Earth and the Moon, for just as the Moon moves about the common centre of mass under the attraction of the Earth, so does the Earth move under the attraction of the Moon. The term $(\mu a/R^2)\cos\psi$ therefore adds nothing new to our knowledge of the potential at the surface of the Earth.

It is the third term that gives the variation of gravity at the surface of the Earth due to the varying position of the Moon. There are of course further terms in the expansion of the potential, but they will be ignored because they are less than the third term by a factor (a/R) or worse; since $a = 6400$ km and $R = 380\,000$ km, the ratio is $1/60$ or less and can be entirely ignored.

To understand the variation of gravity at the surface with the rotation of the Earth and the varying position of the Moon, the angle ψ must be written in terms of the angles that define the directions of P and of the Moon. Take a system of spherical polar co-ordinates with the origin at the centre of the Earth (fig. 43) and with co-latitudes measured from the direction of the north pole of the Earth and with longitudes measured from some direction fixed in space, for example, the first point of Aries (Chapter 4). In these co-ordinates, let θ be the co-latitude of P and θ_M that of the Moon, and let ϕ, ϕ_M be respectively the longitudes of P and of the Moon. Then

$$\cos\psi = \cos\theta\cos\theta_M+\sin\theta\sin\theta_M\cos(\phi-\phi_M)$$

and

$$
\begin{aligned}
3\cos^2\psi - 1 &= 3\cos^2\theta\cos^2\theta_{\mathrm{M}} + \tfrac{3}{2}\sin^2\theta\sin^2\theta_{\mathrm{M}}\{\cos 2(\phi-\phi_{\mathrm{M}})+1\} \\
&\quad + \tfrac{3}{2}\sin 2\theta\sin 2\theta_{\mathrm{M}}\cos(\phi-\phi_{\mathrm{M}}) - 1 \\
&= 3\cos^2\theta\cos^2\theta_{\mathrm{M}} + \tfrac{3}{2}(1-\cos^2\theta)(1-\cos^2\theta_{\mathrm{M}}) - 1 \\
&\quad + \tfrac{3}{2}\sin^2\theta\sin^2\theta_{\mathrm{M}}\cos 2(\phi-\phi_{\mathrm{M}}) \\
&\quad + \tfrac{3}{2}\sin 2\theta\sin 2\theta_{\mathrm{M}}\cos(\phi-\phi_{\mathrm{M}}).
\end{aligned}
$$

Hence

$$
\begin{aligned}
\tfrac{1}{2}(3\cos^2\psi - 1) &= \left(\frac{3\cos^2\theta - 1}{2}\right)\left(\frac{3\cos^2\theta_{\mathrm{M}} - 1}{2}\right) \\
&\quad + \tfrac{3}{4}\sin^2\theta\sin^2\theta_{\mathrm{M}}\cos 2(\phi-\phi_{\mathrm{M}}) \\
&\quad + \tfrac{3}{4}\sin 2\theta\sin 2\theta_{\mathrm{M}}\cos(\phi-\phi_{\mathrm{M}}).
\end{aligned}
$$

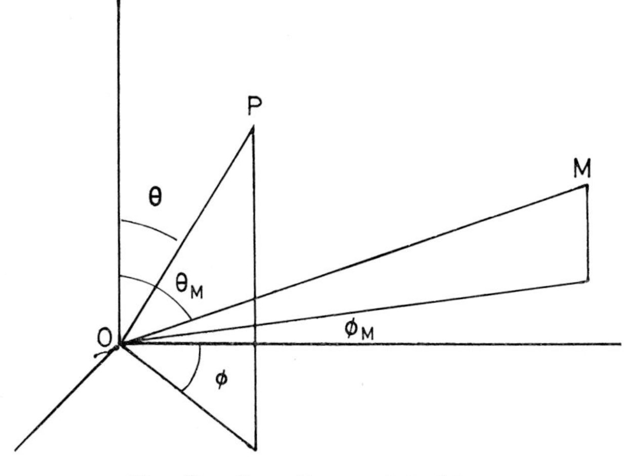

Fig. 43. Co-ordinates of the Moon.

The tides

The quantities in this expression are of three types. First, there is θ, the latitude of the point P, which is a constant for a given point on the surface. Secondly, there is ϕ, the longitude of P, which increases by one revolution (360°) per day irrespective of the position of P; and finally, there are the angular co-ordinates of the Moon that vary in the course of a month with the position of the Moon in her orbit about the Earth. Corresponding to these three sets of quantity, there are three periodic variations in the potential

$$\frac{\mu a^2}{R^3}\tfrac{1}{2}(3\cos^2\psi - 1),$$

the *tide raising potential* as it is called, because it gives rise to the ocean tides.

The term $\frac{3}{4}\sin^2\theta\sin^2\theta_M\cos 2(\phi-\phi_M)$ undergoes a complete variation every time that the angle $2(\phi-\phi_M)$ changes by 360°, that is, twice in the course of a day, and the tide that it produces is therefore known as the *semi-diurnal tide*. Similarly, the term $\frac{3}{4}\sin 2\theta\sin 2\theta_M\cos(\phi-\phi_M)$ undergoes a complete variation once a day, and the tide it produces is known as the *diurnal tide*. It is not strictly correct to say that these terms have periods of exactly 12 and 24 hours because ϕ_M and θ_M are changing as well as ϕ, though much more slowly. In consequence, in addition to the periods of exactly 12 and 24 hours, there are others close to, but not exactly, 12 and 24 hours.

Lastly, there is the term $\{\frac{1}{2}(3\cos^2\theta-1\}\{\frac{1}{2}(3\cos^2\theta_M-1)\}$ which undergoes a complete variation twice in the period of the Moon's revolution in her orbit, that is with a period of a fortnight, and there is an additional variation with the same period arising from the fact that the distance of the Moon, R, is not constant but varies slightly in the course of a month because the orbit of the Moon is slightly elliptical.

The Sun also produces a tide raising potential, but it is smaller in the ratio

$$\left(\frac{\text{mass of Sun}}{\text{mass of Moon}}\right)\cdot\left(\frac{\text{distance of Moon}}{\text{distance of Sun}}\right)^3,$$

that is about 0·42 : 1.

The Sun produces diurnal and semi-diurnal tides just as the Moon does but instead of a fortnightly tide, there is a solar one with a period of half a year.

In the open oceans, the surface of the sea remains an equipotential surface of the combined attraction of the Earth and the Moon and if the solid Earth were rigid and did not yield under the tidal attraction, the height of the sea would follow the tide raising potential, varying with the diurnal, semi-diurnal and longer periods of the potential. Since the potential due to the attraction of the Earth at the radius a is GE/a where E is the mass of the Earth, and since the tide raising potential is of order GMa^2/R^3, the change in radius of the ocean surface would be of order :

$$a\cdot\frac{GMa^2}{R^3}\frac{a}{GE}=\frac{aM}{E}\left(\frac{a}{R}\right)^3.$$

Now M/E is about 1/81 and a/R is about 1/60 and so the oscillations of the sea-level surface would have an amplitude of some 0·4 m. This would be the amplitude of the so-called equilibrium tide ; the tides we observe at the shores of the oceans are much greater because the amplitudes of the tidal currents increase as they flow into shallower water. The relative amplitudes of the diurnal and semi-diurnal tides are also altered by the interference of the tidal currents in shallow seas.

If the Earth were solid and unyielding, the effect of the tide raising potential would be that the value of gravity at a point on the solid Earth would change by

$$\frac{\partial}{\partial a}\left[\frac{\mu a^2}{R^3}\cdot\frac{3\cos^2\psi-1}{2}\right]$$

along the radial direction and by

$$\frac{1}{a}\frac{\partial}{\partial\theta}\left[\frac{\mu a^2}{R^3}\cdot\frac{3\cos^2\psi-1}{2}\right] \text{ and } \frac{1}{a\sin\theta}\frac{\partial}{\partial\phi}\left[\frac{\mu a^2}{R^3}\frac{3\cos^2\phi-1}{2}\right]$$

in the respective directions of the meridian and parallels of latitude. These quantities are of order 10^{-7} of gravity, that is, they are negligible for many purposes but are readily detectable by modern gravity meters.

Tides on the elastic Earth

The actual changes of sea level and gravity are somewhat different because the Earth yields slightly under the attraction of the Sun and Moon. If the tide raising potential is called V_T, the force in a direction x exerted on unit mass at some position in the Earth is $-\sigma(\partial V_T/\partial x)$, where σ is the density of matter in that place. Since this force varies from place to place, the material of the Earth is subject to a stress $(\partial/\partial y)\,.\,\sigma(\partial V_T/\partial x)$ where x, y are co-ordinate directions in the Earth. Such stresses are resisted by the elastic properties of the solid matter of the Earth and it is possible to write the resulting displacements of the surface of the Earth as :

radially : $h\,V_T/g$,

tangentially ; along a meridian : $\frac{l}{g}\frac{\partial V_T}{\partial\theta}$,

along a parallel of latitude : $\frac{l}{g}\frac{\partial V_T}{\sin\theta\,\partial\phi}$,

h and l are known as Love's numbers, after the mathematician A. E. H. Love, who studied mathematically the elastic yielding of the Earth under the attraction of the Sun and the Moon. h and l can be calculated for models of the Earth which give the variation of the density and elastic moduli with distance from the centre of the Earth.

The changing shape of the Earth under the tidal forces corresponds to a redistribution of the material of the Earth that gives rise to a change in the potential at the surface equal to kV_T, where k is a further Love's number. The total change in the potential of gravity at the surface as a result of the combined effects of the original tide raising potential, the redistribution of matter within the Earth and the deformation of the surface, is :

$$V_{\mathrm{T}}+kV_{\mathrm{T}}-\left(\frac{hV_{\mathrm{T}}}{g}\right)g.$$

The factor $(1+k-h)$, usually written as γ, is the factor by which the observed equilibrium tide is reduced below that to be expected for an unyielding Earth ; it is about 0·6. It will be recalled that the actual variation of potential at any point on the surface will depend on the latitude. The factor $(1+k-h)$ has been found from observations of tides in the open oceans. The change in gravity is similarly not $\partial V_{\mathrm{T}}/\partial a$ but is $(1+h-\frac{3}{2}k)$ times the value for the rigid Earth.

The measurement of the factor $(1+h-\frac{3}{2}k)$ is not entirely straightforward, although the sensitivity of modern gravity meters is adequate. Great care has to be taken to insulate the meter from the effects of temperature and pressure, for these may give rise to reading errors with periods of a day ; the drift of the zero of the meter must also be kept as low as possible by careful attention to details of the operation of the apparatus, so that records may be taken over periods of many weeks. The problem is not, however, solved even if sufficiently reliable measurements of gravity can be made, because the value of gravity is affected by the attraction of the waters of the oceans which move under the influence of the tides, and it is difficult to calculate this attraction on account of the poor knowledge of the tidal currents. Evidently, the disturbances will be worse close to the seas and the most reliable measurements of the tidal variation of gravity should be made in the centres of continents. Even then, the results at different places are not in close agreement, as they should be if the Earth were a body with properties that varied only with radius.

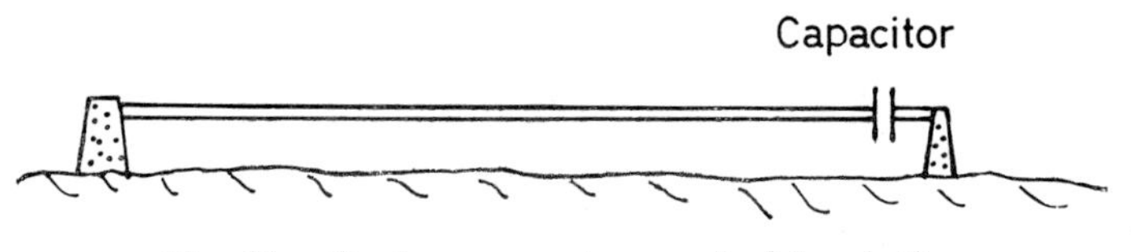

Fig. 44. Strain gauge using a rod of fused silica.

The components of the tidal force tangential to the surface give rise to a stretching of the surface that can be measured, enabling the number l to be found. Again, the effects are very small (1 part in 10^8) and the measurements very delicate. One instrument is shown in fig. 44. A rod of fused silica, chosen because its coefficient of linear expansion is very low and because its length is very stable, is fixed to the ground at one end, while the other end is free. The movement of the ground at this end relative to the rod causes a change in the electrical capacitance between two electrodes fixed respectively to the end of the rod and to the ground. Since it is possible to detect electrically very small changes of capacitance, a change of strain in the ground corresponding to as little as one part in 10^9 of the length of the rod can be detected. Since the length of the rod, against which the movement of the ground is compared, changes both with the temperature of the rod, and with the pressure of the air (on account of the general compression of the rod), the instrument must be isolated as far as possible from changes in the atmospheric conditions, but even so it is not really possible to observe the effects of the fortnightly tides. The development of the gas laser has made possible a strain gauge with which the long period tides may be observed. The width of the spectrum of the radiation emitted by a helium–neon gas laser is so small that it is possible to observe fringes with a Michelson interferometer over path differences of more than 1 km. An interferometer

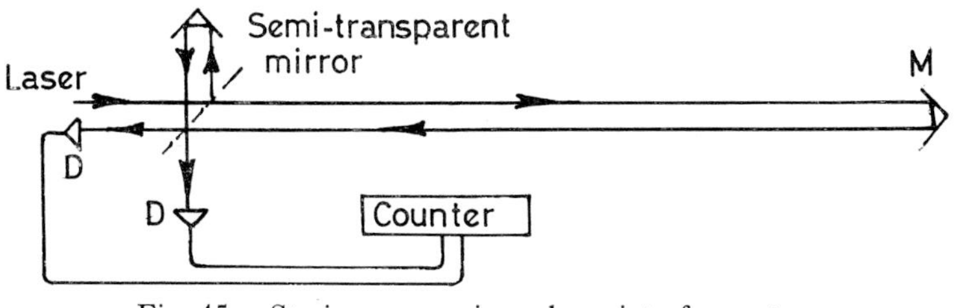

Fig. 45. Strain gauge using a laser interferometer.
D: detector

using this property to measure the changes of strain of the surface of the Earth is shown in fig. 45. A beam of light from the laser is partly reflected at the semi-reflecting mirror and partly transmitted so that one part goes to the mirror close to the semi-reflector and the other part goes to the mirror *M* which may be a few hundred metres away. For ease of adjustment, the mirrors are cube-corner or cat's-eye reflectors (see Chapter 2). After recombination at the semi-reflecting mirror, the two beams fall on two detectors, *D*, the outputs of which are applied to counters that add or subtract counts of fringes according to whether the distance of the mirror *M* is increasing or decreasing. Thus the interferometer measures in terms of interference fringes the increase or decrease of that distance. Three such interferometers have already been set up, two in the U.S.A. and one in Britain, and have shown the diurnal and semi-diurnal stretching and contraction of the surface of the ground and have also shown small oscillations due to waves from distant earthquakes. The tides of long period have not so far been looked for, because to detect them, the wavelength of the laser, in terms of which the strains are measured, must be constan-to a few parts in 10^9 over some months and so far it has not been demont strated that lasers can operate with that stability. However, there is little doubt that they will be made to do so and that the laser interferometer will show the effects of the long period tides.

Because the gravitational attraction of the Sun and the Moon has components tangential to the surface of the Earth, that is perpendicular to the direction of the attraction of the Earth, the direction of the resultant force of the combined attraction varies with the positions of the Sun and Moon with diurnal and semi-diurnal periods. In principle, this effect could be detected by observing the direction of the plumb-line relative to the fixed stars, but in practice that would be ineffective because the directions of the stars cannot be observed accurately enough. The direction of the vertical is therefore observed relative to the surface of the ground but then the measurements do not give directly the change of direction of gravity because the

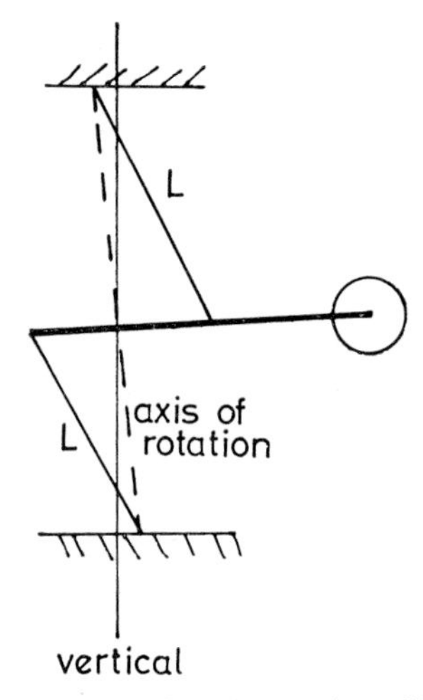

Fig. 46. Tilt meter for observation of Earth tides.

combined effects of the radial expansion and contraction of the Earth under the influence of the tides together with the tangential strains give rise to tilting of the surface. The observed change in direction of the vertical is the difference between the actual change in the direction of gravity and the tilt of the surface, and from it, it is theoretically possible to derive the factor γ. One instrument for measuring tilts is shown in fig. 46. A beam with a heavy mass at one end is supported so that it can rotate about the nearly vertical axis defined by the points at which the ligaments L and L' are attached to the ground. Such a pendulum is very sensitive to changes in the direction of gravity relative to the ground and swings through relatively large angles for small changes in the vertical. It may be arranged to record the changes automatically. The results are even more difficult to interpret than those of gravity measurements, for not only is the horizontal component of gravity more influenced by the attraction of water moving in nearby oceans but also it seems that the tilts of the ground are greatly influenced by local geological structure.

All the methods so far described for measuring the response of the Earth to the tide raising potential measure quite local behaviour, influenced as has been said by the local geological structure, whereas the theories that can be worked out consider the elasticity and density of the Earth as a whole. Recent technical developments have made it possible to determine a value for the number k that does seem to represent the response of the Earth as a whole unconfused by the effects of local structure. It was seen that as a result of the elastic yielding of the Earth, the distribution of mass changes ; there must accordingly be a change in the moment of inertia of the Earth about the polar axis. Since there are no couples acting on the Earth, the spin energy of the Earth is a constant and so the spin angular velocity must change to keep the product $\frac{1}{2}C\omega^2$ constant. Careful comparisons of the times of passage of stars with the times given by atomic clocks show that the speed of rotation of the Earth varies with a period of a fortnight and from that variation the number k can be calculated. The same redistribution of matter causes a change in the external potential and therefore on the forces acting on artificial satellites close to the Earth. The corresponding changes in the orbits of satellites have been detected by very careful observations and again it has proved possible to determine k. The values of k found by these two methods are in good agreement and k must be close to 0·3.

The observed values of Love's numbers are not possible if the Earth is solid throughout and to account for them, it is necessary to suppose that there is a liquid core. With that accepted, the values of the numbers provide a further check on the densities and elastic moduli derived from observations of the times of travel of earthquake waves, but as will be realized from what has been said, they are not at present very stringent checks.

Long-term changes of gravity

The tidal variations of gravity are not the only ones that can be shown to occur, for if the difference of gravity between two sites is measured at frequent intervals over a long period, it is sometimes found to undergo small steady changes. By studying which differences change in some region and which stay constant, it may be possible to show that the values of gravity at some places are not altering while those at others do change. Such measurements have been made in particular around the shores of the Baltic Sea and it seems that there are changes of gravity that are connected with the general subsidence known to be going on in that region. Evidently if the level of the ground is changing relative to sea level, the value of gravity would change simply because the ground level will lie at a different potential. Additional changes may occur if there are movements of mass below the surface, for example, when there is a movement of magma in a volcanic eruption. There is some slight evidence that the direction of gravity at the site of the old Royal Observatory at Greenwich has undergone a very small change over the years with the removal of water from the pores of the chalk in the Artesian basin under London. Such changes are very difficult to detect because they depend on being able to make delicate observations without change of instrumental zero over some years and so far instruments are not good enough for any results to be unambiguous.

While the foregoing possible changes are connected with rather local geological structure, there is also the possibility of an overall change of gravity expressing a change in the gravitational mass of the Earth, in particular as a result of a change in the constant of gravitation. Such a change could only be established through very accurate absolute determinations of gravity. The measurements of Sakuma (Chapter 2) may be accurate enough to detect changes although Sakuma has claimed that in the course of a year, gravity at Sèvres has not changed by more than one part in 10^8. Observations would need to be made at widely distributed stations before one could be sure that any observed change was of general significance and not due to local geological changes, but at the same time, the interpretation to be put on a general change is none too clear. In an absolute determination, one measures the value of gravity in terms of atomic standards of frequency and wavelength and an apparent general change in gravity would mean that the constant of gravitation or the mass of the Earth had altered relative to atomic standards, a result which would certainly be of wide significance for physics although it is by no means evident whether one would be more inclined to believe that the atomic standards or the mechanical or cosmological units had changed. In fact, such questions are experimentally meaningless because one cannot show experimentally that one or the other unit or standard has changed but only that the relation between them has altered.

To summarize this chapter, the value of gravity at the surface of the Earth changes periodically under the tidal influence of the Sun and the Moon, both in magnitude and direction, and the observed changes, together with related effects, provide evidence about the mechanical state of the interior of the Earth. There is some evidence, not too strong, for steady changes of gravity brought about by geological changes and there is the possibility that there might be a general change of gravity related to the general condition of the Universe, although the evidence at present, scant as it is, is against any such change.

Problems

6.1. Given that $h = 0{\cdot}48$ and $k = 0{\cdot}19$, calculate the amplitude of the lunar semi-diurnal variation of g at latitude $50°$, assuming the Moon to be on the equator.

Take the mass of the Moon to be 1/81 of the mass of the Earth, the distance of the Moon to be $3{\cdot}8 \times 10^5$ km and the radius of the Earth to be 6400 km. (0·42 mgal.)

6.2. It has been suggested (by P. A. M. Dirac) that G is proportional to the age of the Universe. Taking that to be 10^{11} years, by how much would you expect the mean value of gravity at the surface of the Earth to change in 10 years? (0·001 mgal.)

Note to problem 6.1

$$\delta g = (1+h-\tfrac{3}{2}k)\left(\frac{\partial V_{\mathrm{T}}}{\partial r}\right)_{r=a},$$

where

$$V_{\mathrm{T}} = \frac{\mu r^2}{R^3}\,\tfrac{3}{4}\sin^2\theta\sin^2\theta_{\mathrm{M}}\cos 2(\phi-\phi_{\mathrm{M}}).$$

So

$$\left(\frac{\partial V_{\mathrm{T}}}{\partial r}\right)_{r=a} = \tfrac{3}{2}\frac{\mu a}{R^3}\sin^2\theta\sin^2\theta_{\mathrm{M}}\cos 2(\phi-\phi_{\mathrm{M}}).$$

Remember that μ is $G \times$ mass of Moon (M).

Divide by $g = \dfrac{G}{a^2} \times$ mass of Earth (E):

$$\frac{1}{g}\left(\frac{\partial V_{\mathrm{T}}}{\partial r}\right)_{r=a} = \tfrac{3}{2}\frac{M}{E}\left(\frac{a}{R}\right)^3\sin^2\theta\ldots$$

Remember that $\theta = \tfrac{1}{2}\pi -$ latitude.

CHAPTER 7

the constant of gravitation

History of measurements

THE constant of gravitation was the first of the so-called fundamental constants of physics to be measured in an experiment on Earth, for although Römer had earlier estimated the speed of light from observations of the satellites of Jupiter, Cavendish's experiments on the constant of gravitation antedated experiments on the speed of light. With such a long history it seems at first sight surprising that our present knowledge of the constant of gravitation is little better than that given by the work of Cavendish. The reason is that the force of gravity is really very weak and in any experiment to measure it, disturbances from other sources compete for attention, so that technical advances that enable more accurate observations to be made are not of much help when the effects to be observed are not wholly those of gravitation. Apart from this inherent difficulty of the determination of the constant, there has not been until recently any great requirement for a knowledge of it, since for most purposes we do not need to know the actual value. Orbits of satellites, the shape and size of the Earth, geophysical applications of gravity measurements, the use of known values of gravity in physics, all are independent of a knowledge of the constant of gravitation, and it is only recently that it has become interesting to know the actual value. One reason is that it would be of great interest to our general ideas of physics to know if the ' constant ' is not constant but changes in time, and that can only be found from measurements of the constant in terms of our terrestrial units. Other aspects of gravity, whether there is shielding, whether the constant depends on the type of material, can be investigated, as will be explained, without measurements of the constant itself. The second reason is a geophysical one. The studies of the interior of the Earth that have been mentioned in earlier chapters have in recent years been so thoroughly developed that it is now possible to say what is the equation of state of matter in certain zones of the Earth, that is to say, what is the relation between the density of the material and the temperature and pressure, and so it should be possible to compare such relations with those determined in the laboratory. However, masses in the two relations are not the same, because in the laboratory they are found in terms of the laboratory mass standard, the kilogramme, whereas in the interior of the Earth they are derived from gravity measurements combined with a value for the constant of gravitation, and a value of G is therefore needed to make the comparison.

Cavendish used the torsion balance to determine G and following his work, many improvements were made up to the beginning of the twentieth century. The principle of the experiment is shown in fig. 47. A beam of length $2l$ is suspended from a torsion fibre of length h and carries masses m at each end. Large attracting masses

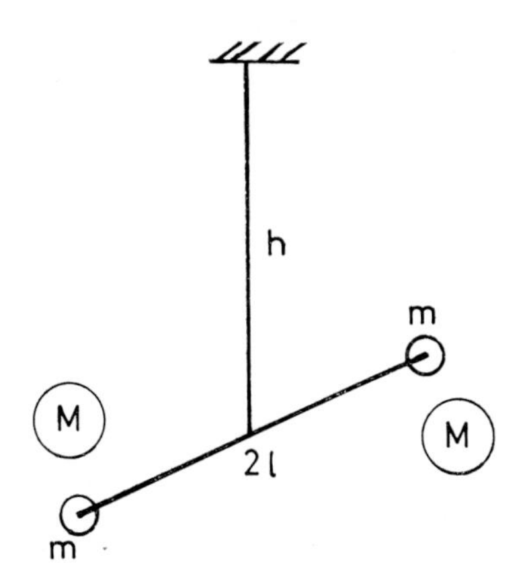

Fig. 47. Torsion balance determination of G.

M are placed close to the ends of the beam. Let the beam twist through an angle θ when the large attracting masses are brought into position. If the torsional constant of the fibre is τ, the couple $\tau\theta$ is balanced by the gravitational couple between the masses, that is:

$$\tau\theta = \frac{2GMml}{d^2}.$$

If the beam is allowed to swing freely with the large masses removed, its period of oscillation will be:

$$T = 2\pi\left(\frac{ml^2}{\tau}\right)^{\frac{1}{2}}$$

and so

$$G = 2\pi\frac{\theta l d^2}{MT^2}.$$

Many difficulties afflict this experiment. θ is larger the larger the free period of the beam, but the larger T the longer it takes for the beam to come into equilibrium and so the more difficult it is to measure θ. At the same time, the beam responds readily to any other steady couples exerted on it, in particular, any due to convection currents in the surrounding air, although those can be eliminated by putting the apparatus in a vacuum. C. V. Boys, who studied this experiment very thoroughly, showed that the best results were obtained by making the apparatus small and by using a very fine torsion fibre—he developed very fine fibres of fused silica for the purpose.

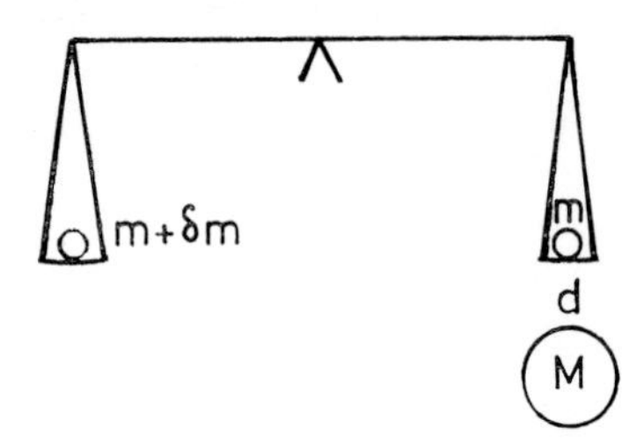

Fig. 48. Chemical balance determination of G.

The constant of gravitation can also be found from measurements with a chemical balance, as shown in fig. 48. Let the balance be initially in equilibrium with masses m in each pan and let a mass M be placed under one pan at a distance d below it. The force it exerts on the mass m above it is :

$$\frac{GMm}{d^2}$$

and if a mass δm has to be added to the other pan to restore the balance to equilibrium :

$$g\delta m = \frac{GMm}{d^2}.$$

A new method of measurement

The simple torsion balance experiment and the chemical balance both suffer from the disadvantage that very small angles have to be measured—the rotation of the torsion beam or the out-of-balance rotation of the beam of the chemical balance—and they have to be measured on a system with a very long period. These are unfavourable conditions for measurements of precision, in particular, it is difficult to reduce the scatter due to random errors of measurement by repeating the observations many times under the same conditions. A considerable improvement should be possible if G is found from the period of the torsional pendulum oscillating in the field of the gravitational attraction of the large masses. The principle of the method is indicated in fig. 49. Two experiments are done in one of which the large masses are placed in positions in line with the beam of the pendulum while in the second they are placed in positions at right angles to the beam. In the first position the gravitational attraction

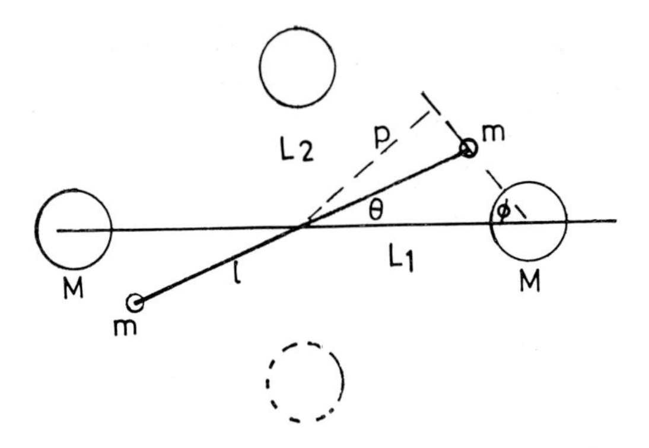

Fig. 49. New method for the determination of G.

increases the restoring force and the period of oscillation of the beam is less than free period, while in the second the gravitational attraction acts against the restoring force of the torsional fibre and the period is greater. In the first position, the force between the nearby masses m and M when the beam is displaced through an angle θ from the equilibrium position is :

$$\frac{GMm}{D^2} = \frac{GMm}{L_1{}^2 + l^2 - 2L_1\, l \cos\theta},$$

L_1 and l being as indicated in fig. 49 ; and the couple exerted on the beam by the two pairs of masses is :

$$2p\,\frac{GMm}{D^2} = 2L_1 \sin\phi \frac{GMm}{D^2} = \frac{2GMmL_1\, l \sin\theta}{(L^2 + l^2 - 2L_1\, l \cos\theta)^{3/2}}.$$

The total couple acting on the beam is the sum of this gravitational couple on the small masses, plus the torsional couple exerted by the suspension, $\tau\theta$, say, plus the gravitational couple exerted by the large masses on the structure of the beam, which will be written as $B_1\,\theta$. Ignoring θ^2 and higher powers, the total couple acting on the beam will be :

$$\tau\theta + B_1\,\theta + \frac{2GMmL_1\, l\,\theta}{(L_1 - l)^3}.$$

This couple must balance the product of the moment of inertia and the angular acceleration of the beam, that is :

$$(I + 2ml^2)\,\ddot{\theta},$$

where I is the moment of inertia without the small masses and $2ml^2$ is the moment of inertia of the two small masses. Ignoring the attraction of each mass M on the small mass further from it (which cannot in practice be ignored) the equation of motion of the beam for small angles θ is

$$(I+2ml^2)\ddot{\theta}+\left(\tau+B_1+\frac{2GMmL_1l}{(L_1-l)^3}\right)\theta = 0.$$

This equation may be written as :

$$(I+2ml^2)\,\ddot{\theta}+\{\tau+B_1+GMmF_1(L_1,\,l)\}\theta = 0.$$

$F_1(L_1, l)$ is equal to $2L_1\,l/(L_1-l)^3$ and is a geometrical factor depending on L_1 and l alone. The equation represents a simple harmonic motion with an angular frequency given by :

$$(I+2ml^2)\,\omega_1^2 = \tau+B_1+GMmF_1(L_1,\,l).$$

When the large masses are in the position at right angles to the beam, the attraction reduces the restoring couple, the period is lengthened and the angular frequency is given by :

$$(I+2ml^2)\,\omega_2^2 = \tau-B_2-GMmF_2(L_2,\,l),$$

where $F_2(L_2, l)$ is another geometrical factor, and B_2 is the couple exerted on the structure of the beam by the large masses in this position.

The equations of motion in both positions involve the moment of inertia of the beam and the attractions of the large masses on the structure of the beam, both of which may be difficult to calculate or determine if the beam is not of the very simplest form. By moving the small masses to a second position, I, B_1 and B_2 may be eliminated. Let the new distance of the small masses from the centre of the beam be l'. The equations for the periods are then :

$$(I+2ml'^2)\,\omega_3^2 = \tau+B_1+GMmF_1(L_1,\,l'),$$
$$(I+2ml'^2)\,\omega_4^2 = \tau-B_2-GMmF_2(L_2,\,l').$$

Putting

$$F_1(L_1,\,l') = F_1',\;\; F_2(L_2,\,l') = F_2',$$

it follows that

$$G = \frac{2(l^2-l'^2)(\omega_2^2\omega_3^2-\omega_1^2\omega_4^2)}{M\{(F_1-F_1')(\omega_2^2-\omega_4^2)+(F_2-F_2')(\omega_1^2-\omega_3^2)\}}.$$

In this equation all the geometrical factors can be measured or calculated with high accuracy and the precision of the determination of G depends entirely on that of the measurement of the periods.

No experiment has yet been performed according to this scheme, but Heyl and his colleagues at the American National Bureau of Standards made determinations with the small masses in just the one position on the beam. The moment of inertia I does not have to be known but the factors B_1 and B_2 that give the attraction of the large masses on the structure of the beam do have to be calculated. A new determination according to the scheme in which B_1 and B_2 are eliminated is now being prepared by the British National Physical Laboratory and the University of Trieste. The experiment will be performed in the large cave, the Grotta Gigante, near Trieste, enabling the torsional suspension to be some 80 m long. As a consequence, the torsional couple can be made less than the gravitational couple in the 'in-line' position and only slightly greater than the gravitational couple in the 'right angles' position, and so the differences between the periods in the two positions should be quite large. The large masses will weigh some 500 kg, and will be made in sections. Evidently the position of the centre of mass of each large mass must be accurately known, and that involves not only accurate measurement of the dimensions but also knowing that the density of the material is uniform. If the large masses were each to be made in one piece, the uniformity of the density could not be checked; instead, the masses are made from a set of discs stacked to form cylinders, and by checking the dimensions, mass and the position of the centre of mass of each component disc, the position of the centre of mass of the stack should be accurately related to the external dimensions of the stack. The periods of the beam in the various positions will be of the order of 1000 to 2000 s and will be found from continuous automatic recording of the angular position of the beam followed by an analysis of the records in a computer.

The results of a number of determinations of G are collected in Table 3. Evidently they are not in good agreement and there is no

Table 3. Determinations of the constant of gravitation

Author		Method	G $\mathrm{Nm^2/kg^2}$
H. Cavendish	1798	Torsion balance, deflection	$6{\cdot}754 \times 10^{-11}$
J. H. Poynting	1891	Chemical balance	6·698
C. V. Boys	1895	Torsion balance, deflection	6·658
K. Braun	1896	Torsion balance, deflection	6·658
P. R. Heyl	1930	Torsion balance, period	6·670

possibility of using direct measurements of G to investigate any dependence of G on materials, any shielding effect or any variation of G with time. The first two possibilities can be studied by very delicate experiments that are designed to detect small changes of G without actually measuring it.

Is the constant constant?

The independence of G from the nature of the materials was established to a very high accuracy by the Hungarian Baron Eötvös, using the apparatus shown in fig. 50. A torsion balance is provided with masses at the end of the beam which have the same mass as determined by balancing on a chemical balance but which are of different materials. The beam is allowed to come to rest in an east–west direction and then the support of the torsional suspension is turned through 180°; the new rest position of the beam is observed. If the

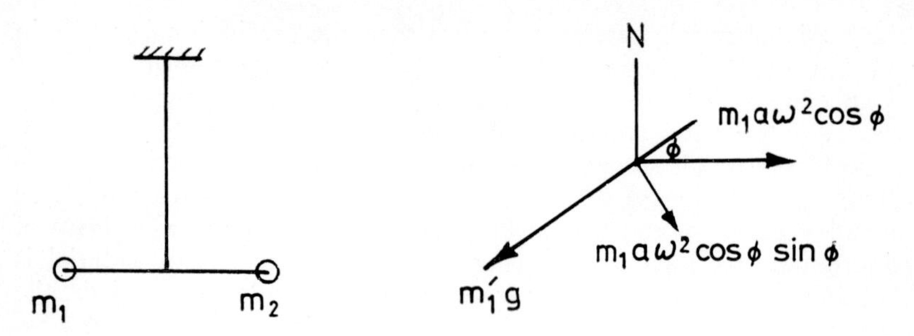

Fig. 50. Eötvös torsion balance experiment on G.

angle between the two rest positions differs from 180°, a dependence of gravity on material would be indicated.

As shown by the vector diagram, the forces acting on one of the masses are $m_1'g$ towards the centre of the Earth and $m_1 a\,\omega^2 \cos\phi$ perpendicular to the axis of rotation of the Earth; m_1 is the inertial mass which gives the force corresponding to an acceleration, and m_1' is the gravitational mass giving the attraction under gravity. a is the radius of the Earth, ϕ the latitude, and ω the Earth's spin angular velocity.

The horizontal component of the force on the mass is:

$$m_1\, a\omega^2 \cos\phi \sin\phi,$$

and the couple on the beam due to a possible difference between the masses at either end is:

$$l(m_1 - m_2)a\omega^2 \cos\phi \sin\phi,$$

where l is the length of the beam.

If the torsional constant of the suspension is τ, the angle through which the beam turns would differ from 180° by:

$$\frac{1}{\tau}\, 2l(m_1 - m_2)\, a\omega^2 \cos\phi \sin\phi.$$

Now the resultant force on one of the masses is:

$$(m'^2_1 g^2 + m_1^2 a^2\omega^4 \cos^2\phi)^{\frac{1}{2}}$$

and the condition for the two masses to balance on a chemical balance is:

$$m'^2_1\, g^2 + m_1^2\, a^2\omega^4 \cos^2\phi = m'^2_2 g^2 + m_2^2\, a^2\, \omega^4 \cos^2\phi.$$

If

$$m' = m + \delta m,$$

$$m_1^2/m_2^2 = 1 + 2(\delta m_1 - \delta m_2)$$

and hence

$$m_1 - m_2 = \frac{m_1^2 - m_2^2}{m_1 + m_2} \doteqdot \delta m_1 - \delta m_2.$$

Eötvös was able to make a torsion balance sensitive enough that he could show that the gravitational masses of different materials differed by less than 1 part in 10^9 when the inertial masses were equal.

It might seem at first sight that the forces exerted by the Sun on the pendulum should be allowed for. (Since we are concerned with the direct attraction of the heavenly body on the pendulum, an attraction proportional to M/d^2, the effect of the Sun is 200 times greater than that of the Moon.) Now when the Sun is rising or setting, its attraction is along the direction of the beam and if there is any differential force it will not produce any couple. On the other hand, at midday or midnight, the Sun is at right angles to the beam and a differential attraction would give a net couple. Observations made over a long time or at corresponding solar times would eliminate the effect of the Sun.

However, the effect of the Sun can be put to use, for by setting the beam in the north–south instead of in the east–west direction, neither the gravitational attraction of the Earth nor the acceleration corresponding to the spin angular velocity of the Earth can produce net couples and the attraction of the Sun on the two masses may be compared with the accelerations due to the rotation of the masses around the Sun as they are carried by the Earth in its orbit.

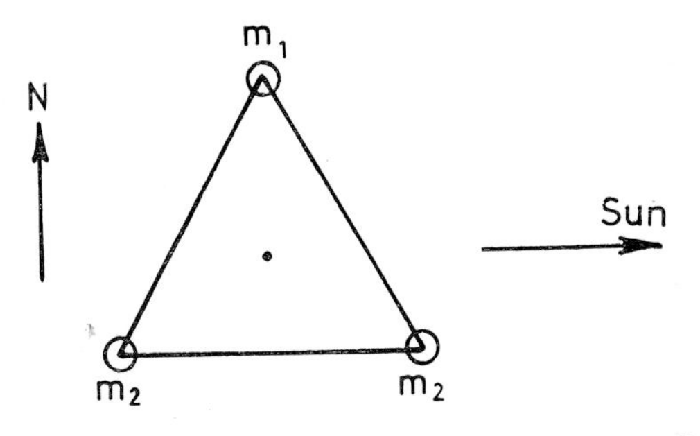

Fig. 51. Dicke torsion balance experiment on G.

Professor R. H. Dicke of Princeton University has recently performed a refined experiment based on this principle. The apparatus is shown in fig. 51. The torsional pendulum consists of three masses placed at the corners of a frame in the form of an equilateral triangle. Two of the masses are of the same material, the third is different. The horizontal forces on the masses comprise the inertial forces due to rotation about the Sun, which are assumed to be all the same, and the attractions of the Sun, which may differ with the material of the masses.

Let p_i be the perpendicular distance from the centre of mass of the beam to the line joining the ith mass to the Sun. Let the m_i be the inertial mass and a_I the acceleration of the ith mass corresponding to its orbital motion about the Sun. Similarly let m_i' be the gravitational mass and a_G the gravitational acceleration. The net couple the ith mass exerts on the beam is then :

$$p_i(m_i' a_G + m_i a_I),$$

and the total couple exerted by all three masses is

$$\sum m_i p_i \left\{ \frac{m_i'}{m_i} a_G + a_I \right\}.$$

If all three masses are of the same material, $m_i' = m'$, $m_i = m$, say, there is no net torque, and so

$$\left(m \left(\frac{m'}{m} \right) a_G + a_I \right) \sum p_i = 0,$$

which is satisfied if the perpendiculars p_i are measured from the centre of mass of the beam.

If one mass B differs from the other two, A, we can write the difference :

$$\left(\frac{m'}{m} \right)_B - \left(\frac{m'}{m} \right)_A \quad \text{as} \quad \eta(A, B).$$

Then we see that the net torque on the balance beam for such a system is just :

$$m_B p_B a_G \eta(A, B).$$

Now a_G is GM_s/R_s^2 where M_s is the mass of the Sun and R_s the distance of the Sun. The perpendicular distance p_B depends on the direction of the Sun and varies periodically as the direction of the Sun changes in the course of 24 hours. As the position of the Sun changes, the direction of the resultant couple on the pendum also changes and so the pendulum should undergo a periodic change of angular position if the gravitational mass depends on material. Since the period of the effect is known, it is easier to detect than a static change, and Dicke was able to show that the dependence of gravitational attraction on material is less than part in 10^{11} for the materials he used, which were lead chloride and copper in one group of experiments and aluminium and gold in another. He has pointed out that because all atomic nuclei are composed of protons and neutrons in varying proportions, his result means that there is no detectable difference between the gravitational attractions of these two particles.

Delicate experiments have also shown that the presence of matter between two other bodies has no detectable effect on the attraction between the two bodies, that is, there is no gravitational screening. A recent experiment was that done at the University of Trieste using apparatus installed in the Grotta Gigante for the measurement of Earth tides. The apparatus is a horizontal pendulum having a very long period and very high sensitivity, and advantage was taken of a solar eclipse that occurred low on the horizon to the west of Trieste. The Moon lay between the Sun and the pendulum, and the attraction of the Sun on the pendulum would have been reduced if the Moon had screened the attraction of the Sun. The effect would be a discontinuity in the record of the tilt due to the Earth tides at the time of totality. No such effect was observed and the screening effect must be very small if it occurs.

Delicate modern experiments confirm that to a very high degree of accuracy, the gravitational attraction between two bodies is indeed given by the formula of Newton :

$$F = \frac{Gm_1 m_2}{d^2},$$

but experiments are not able to determine whether or not the constant G may change with time relative to our terrestrial standards of measurements. To investigate that question will require extreme experimental skill and great imagination in the design of experiments, and remains a challenge for the future.

Problems

7.1. In a Cavendish experiment, the free period of the torsion balance is 100 s. If the beam is 2 m long and the attracting masses are placed 10 cm from either end, what is the deflection for attracting masses of 100 kg ? $G = 6{\cdot}67 \times 10^{-11}$ Nm²/kg². (10^{-3} rad.)

7.2. The sensitivity of a chemical balance is 2 min of arc per mg. If a mass of 1 kg is placed on each pan and a mass of 500 kg is placed 50 cm below one pan, calculate the deflection of the beam. (1″·5.) $g = 9{\cdot}8$ m/s², $G = 6{\cdot}67 \times 10^{-11}$ Nm²/kg².

FURTHER READING

D. R. Bates (editor), *The Planet Earth* (Pergamon, Oxford, London, 1964).

L. V. Berkner and H. Odishaw (editors), *Science in Space* (McGraw-Hill, New York, 1961).

A. H. Cook, ' The absolute determination of gravity ', *Contemporary Physics*, vol. 8 (1967), pp. 251–266.

A. H. Cook, ' A new determination of the constant of gravitation ', *Contemporary Physics*, vol. 9 (1968), pp. 227–238.

M. B. Dobrin, *Introduction to Geophysical Prospecting* (McGraw-Hill, New York, 1952).

T. F. Gaskell (editor), *The Earth's Mantle* (Academic Press, London, 1967).

J. C. Harrison, ' Measurements of gravity at sea ', in *Methods and Techniques of Geophysics*, edited by S. K. Runcorn (Interscience, London, 1960).

W. M. Kaula, *An Introduction to Planetary Physics* (Wiley, New York, 1968).

D. G. King-Hele, *Observing Earth Satellites* (McMillan, London, 1966).

D. G. King-Hele (organizer), ' A discussion on orbital analysis ', *Philosophical Transactions of the Royal Society*, A, vol. 262 (1967), pp. 1–202.

INDEX

THE WYKEHAM SCIENCE SERIES

for schools and universities

1 *Elementary Science of Metals* — J. W. MARTIN and R. A. HULL
(S.B. No. 85109 010 9)*

2 *Neutron Physics* — G. E. BACON and G. R. NOAKES
(S.B. No. 85109 020 6)*

3 *Essentials of Meteorology* — D. H. MCINTOSH, A. S. THOM and V. T. SAUNDERS
(S.B. No. 85109 040 0)*

4 *Nuclear Fusion* — H. R. HULME and A. MCB. COLLIEU
(S.B. No. 85109 050 8)*

5 *Water Waves* — N. F. BARBER and G. GHEY
(S.B. No. 85109 060 5)*

6 *Gravity and the Earth* — A. H. COOK and V. T. SAUNDERS
(S.B. No. 86109 070 2)*

7 *Relativity and High Energy Physics* — W. G. V. ROSSER and R. K. MCCULLOUGH
(S.B. No. 81509 080 X)*

Price per book for the Science Series **20s.—£1·00 net** *in U.K. only*

THE WYKEHAM TECHNOLOGICAL SERIES

for universities and institutes of technology

1 *Frequency Conversion* — J. THOMSON, W. E. TURK and M. BEESLEY
(S.B. No. 85109 030 3)*

Price per book for the Technological Series **25s.—£1·25 net** *in U.K. only*

* Standard Book Catalogue numbers.

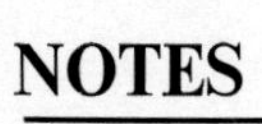

NOTES